Photoinduced Defects in Semiconductors is the first book to give a complete overview of the properties of deep-level, localized defects in semiconductors that can be created or destroyed by light. Such comparatively long-lived (or metastable) defects exhibit complex interactions with the surrounding material and can significantly affect the performance and stability of certain semiconductor devices.

After an introductory discussion of metastable defects in terms of configuration-coordinate diagrams, energy bands, and energy levels, the properties of DX and EL2 centers in III–V compounds are presented. Additional crystalline materials are also treated, and then a detailed description is given of the properties and kinetics of photoinduced defects in hydrogenated amorphous silicon, as well as other amorphous semiconductors. The book closes with an examination of the effects of photoinduced defects in a range of practical applications, such as III–V and II–VI semiconductor devices, xerography, and amorphous silicon solar cells and thin-film transistors.

Throughout, unifying concepts and models are stressed, and the book will be of great use to graduate students and researchers interested in the physics and materials science of semiconductors.

T0275625

Cambridge Studies in Semiconductor Physics
and Microelectronic Engineering: 4

EDITED BY
Haroon Ahmed
Cavendish Laboratory, University of Cambridge
Michael Pepper
Cavendish Laboratory, University of Cambridge
Alec Broers
Department of Engineering, University of Cambridge

PHOTOINDUCED DEFECTS IN SEMICONDUCTORS

TITLES IN THIS SERIES

PHOTOINDUCED DEFECTS IN SEMICONDUCTORS

DAVID REDFIELD

Consulting Professor of Materials Science
Stanford University

and

RICHARD H. BUBE

Professor Emeritus of Materials Science and Electrical Engineering
Stanford University

CAMBRIDGE
UNIVERSITY PRESS

CAMBRIDGE UNIVERSITY PRESS
Cambridge, New York, Melbourne, Madrid, Cape Town, Singapore, São Paulo

Cambridge University Press
The Edinburgh Building, Cambridge CB2 2RU, UK

Published in the United States of America by Cambridge University Press, New York

www.cambridge.org
Information on this title: www.cambridge.org/9780521461962

First published 1996
This digitally printed first paperback version 2006

A catalogue record for this publication is available from the British Library

Library of Congress Cataloguing in Publication data
Redfield, David.
Photoinduced defects in semiconductors / David Redfield, Richard H. Bube.
p.   cm. – (Cambridge studies in semiconductor physics
and microelectronic engineering ; 4)
Includes index.
ISBN 0-521-46196-0 (hc)
1. Semiconductors – Defects.   2. Photochemistry.
I. Bube, Richard H., 1927– .   II. Title.   III. Series
QC611.6.D4R43   1996
621.3815′2 – dc20   95–32567
CIP

ISBN-13 978-0-521-46196-2 hardback
ISBN-10 0-521-46196-0 hardback

ISBN-13 978-0-521-02445-7 paperback
ISBN-10 0-521-02445-5 paperback

To all my teachers. [DR]

Of old Thou didst lay the foundation of the earth,
and the heavens are the work of Thy hands.
They will perish, but Thou dost endure;
they will all wear out like a garment.
Thou changest them like raiment, and they pass away;
But Thou art the same, and Thy years have no end.
(Psalm 102:25–27 RSV) [RHB]

Contents

Preface

Over the past few decades a fascinating revolution has taken place with regard to the way that defects in semiconductors and their properties are regarded. Traditional understandings of a variety of effects have used models of isolated point defects with fixed properties, operating independently in the matrix of the solid, and describable in terms of one-electron energy bands and defect levels. More recently we have come to realize that this classic picture of a simple defect with fixed properties may be the exception – that in real semiconductors there is a wide variety of interactions between defects with deep energy levels and the lattice that leads to variations in local atomic configurations comprising the defect, and in some cases to complexes between defects. The central distinguishing feature of most of the defect processes of interest is that they inextricably couple the electronic system and local structural configurations. Consequently, processes involving these defects cannot be fully described by either electronic transitions or defect states alone.

Some of these new configurations are relatively stable, but many others are not: They can be brought into metastable existence in the solid by the action of photoexcitation, other high-energy excitation, injection of excess carriers, and/or thermal energy, and they can be removed once again by thermal or optical anneal that returns them to their ground state. The metastability of their existence and the radical changes in electronic properties that can exist between their ground state and metastable state provide interesting new possibilities for semiconductors, but often they also provide mechanisms for instability and degradation that handicap the applications of semiconductors. It is our purpose in this book to provide a background for understanding these effects as they occur in a variety of both crystalline and amorphous materials, together with illustrations of the phenomena that demonstrate the nature of the effects observed and similarities among different kinds of defects.

Our research experience with photoinduced defect changes extends back thirty-nine years, but the immediate motivation for preparing this book arises from our extended investigations in the past ten years of the metastable defects in hydrogenated amorphous silicon (a-Si:H). Questions associated with these defects, as with other photoinduced defect processes, are many, but the significance of the results affecting the stability of a-Si:H is both theoretically and practically great. As the work proceeds, striking similarities are found among the properties and behavior of photoinduced and related defects in a variety of II–VI and III–V semiconductors, as well as in amorphous semiconductors.

We hope this book may serve a useful role in integrating the various, apparently unrelated effects involved in defect processes in semiconductors, and in pointing the way to possible future avenues of approach that may prove to be fruitful. The book has been written on the assumption that the reader has a basic knowledge of semiconductor physics, and an introductory chapter describes general properties of the metastable defects that are of current interest. We have chosen to deal with all these topics in a nonmathematical way, emphasizing the concepts instead. That these concepts are exhibited using a generally graphical approach should not be interpreted as signifying a lack of rigorous calculations in the underlying research. There have been many such calculations in the references given, particularly in connection with the DX center. Amorphous silicon, which has been our research emphasis, is less amenable to such rigor, so analogy has played a strong role.

We thank our colleagues, with whom we have discussed these questions, our students, who have inspired us, and our loved ones, who have supported and encouraged us.

1

Introduction: Metastable Defects

1.1 Types of Defects

This discussion of defects in semiconductors deals with those having significant photoelectronic interactions. The word *defect* is used here as a shorthand
for "imperfection." It may therefore include any departure from the ideal periodic lattice of a crystal or, in the case of amorphous materials, any departure
from an ideal continuous random network. These defects may take a variety
of forms:

1. native point defects, such as isolated vacancies, interstitials, or antisite
   atoms of the host crystal;
2. point defects associated with the presence of isolated impurity atoms, in
   either substitutional or interstitial positions;
3. defect complexes formed by the spatial correlations between different
   point defects, such as donor–acceptor or impurity–vacancy pairs;
4. line defects, such as dislocations;
5. defects associated with grain boundaries in a polycrystalline material;
   and
6. defects associated with the existence of a surface or interface.

In keeping with the thrust of this book, some defects (e.g., phonons, dislocations, and interfaces or surfaces) are not treated independently.

1.2 General Effects of Defects on Electronic Properties

Any of the above defects can play a variety of electronically active roles that
affect the electrical and optical properties of a semiconductor. Some of the
traditionally accepted roles can be summarized as follows.

Donor or Acceptor Fundamentally, a *donor* is a defect that is neutral when electron-occupied, or positive when unoccupied; an *acceptor* is a defect that is negative when electron-occupied, or neutral when unoccupied. In many cases this means that a donor defect has an extra electron that it can contribute to the extended states in the conduction band, or an acceptor defect has a deficiency of an electron that can be filled from the valence band, in which it produces a hole. The presence of such defects affects the density of free carriers and hence the conductivity of the material. Important parameters of these defects are their density and electron (donor) or hole (acceptor) ionization energy.

Trap A defect can capture an electron (*electron trap*) or hole (*hole trap*) with such a small thermal ionization release energy that the trapped carrier is in general released thermally to the nearest band before capture of (i.e., recombination with) a carrier of opposite type can occur. A consequence of this rapid interchange is that the occupation probability of such traps can be described by the same Fermi–Dirac function, with the same quasi-Fermi energy, as that of the adjacent band. The presence of these defects can lead to carriers in localized states near the band edge. A typical effect of such trapping is to produce a decay time for photoconductivity, for example, that is longer than the simple lifetime of a free carrier. Important parameters of traps are their density and their ionization energy for the trapped carrier.

Recombination Center A defect can capture an electron or a hole with such a large thermal ionization energy that the captured carrier has a high probability of recombining with a carrier of the opposite type before being thermally re-excited to the band. The presence of recombination centers usually reduces the lifetime of free carriers. Important parameters of recombination centers are their density and their capture cross sections for electrons and holes.

When capture or recombination occurs, the excess energy of the recombining carrier must be released; this release of energy occurs either by the creation of a photon (the radiative process, important to luminescence), by the release of many phonons (a nonradiative process), or by the excitation of free carriers (an Auger process).

Specific names have often been associated with recombination centers that have striking effects:

1. *Sensitizing centers* have a large capture cross section for minority carriers but a much smaller capture cross section for majority carriers, so that the lifetime of majority carriers, and hence the magnitude of the photoconductivity, is greatly increased.

2. *Killer centers* have a large capture cross section for majority carriers, so that the lifetime of majority carriers and the magnitude of the photocon- ductivity are drastically decreased.
3. *Poison centers* have a large nonradiative capture cross section for carri- ers, so that they compete with other defects with a radiative capture cross section and reduce the luminescence efficiency.

Optical Absorption Center   An electron associated with the defect can be photoexcited from the defect to the conduction band, from the valence band to the defect, or between the ground state and excited states of the defect. In this way the defect makes a contribution to the *extrinsic* optical absorption of the semiconductor. Important parameters are the optical cross section S_{opt} and the density of defects available for photoexcitation N_I; $\alpha = S_{opt} N_I$, where α is the corresponding absorption coefficient. Values of S_{opt} are usually of the or- der of 10^{-16} cm^2.

Scattering Center   Because they disturb the periodicity of the perfect lattice, defects behave as scattering centers in determining the mobility of free car- riers. Important parameters are the scattering cross section S_{sc} and the defect density. If the defects are charged, they have a large Coulombic cross section for scattering ($S_{sc} \approx 10^{-12}$ cm^2) and make a major impact on the mobility. If the defects are neutral ($S_{sc} \approx 10^{-16}$ cm^2), their effect on the mobility is much smaller and in general important only at low temperatures and for high densi- ties of neutral defects. When present in high densities, defects also create dis- order potentials that can affect the mobility in ways other than simple "impu- rity" scattering.

Each of these descriptions is functional; that is, the same defect can play the role of donor, electron trap, recombination center for holes, optical ab- sorption center, or scattering center. It is also possible for the same defect to act as a trap under one set of temperature and photoexcitation conditions, and then as a recombination center under another set of conditions.

1.3 The Need for New Models of Defects

In the past twenty years, there have been reports of a number of observations of electronic properties of semiconductors that cannot be explained within the framework of the familiar one-electron energy bands and defect energy levels in the fundamental gap. These observations have challenged our understand- ings and have been linked with impediments to substantial applications of

some of these materials. Much theoretical and experimental effort has gradually produced a number of explanations that are now widely accepted, but the
diversity of the materials and types of phenomena involved have created a
fragmented picture. This book is directed at the integration of these observations and their explanations in several classes of semiconductors.

The central common property of these "new" observations is that they represent enduring departures from equilibrium conditions. In many cases these
departures may be initiated by external excitation, such as by absorbing light,
and the resulting nonequilibrium state of the material endures for significant
times. The state of interest is thus termed *photoinduced,* and it is *metastable.*
Its properties may be studied by a variety of techniques, and its relaxation in
time at different temperatures provides key kinetic information about the
metastable state that is used to develop explanatory models. With hindsight
one could say that kinetic data alone exclude traditional one-electron models
for many (but not all) of these phenomena.

To illustrate the problem we choose one of the earliest and most graphic of
these observations, persistent photoconductivity (PPC) in a homogeneous material (discussed in some detail in later chapters). We show in Figure 1.1 data
on the Hall coefficient R_H in the dark, hence the electron density, as a function of temperature in an *n*-type Te-doped $Al_{0.36}Ga_{0.64}As$ alloy from Nelson
(1977). The solid curve was obtained simply by cooling in the dark, whereas
the upper dashed curve describes a state formed by photoexcitation at low
temperature. The difference between this PPC and ordinary photoconductivity is the persistence at low temperatures of the high-electron-density state after the light is turned off. At the lowest temperatures the decay times become
immeasurably long and incompatible with normal lifetime-limiting recombination processes. Another attribute that was difficult to explain is that decay
kinetics were clearly nonexponential in the intermediate temperature range
63–82 K (Nelson 1977). At room temperature the metastable condition disappears and the process can be repeated.

This report of PPC was not the first, and the particular material involved is
not unique. A variety of possible explanations have been proposed, including
the existence of defects with high Coulomb-repulsive barriers, association of
defect levels with secondary conduction-band minima, or spatial inhomogeneities in the material. (Examples of these are cited in later chapters.) It is
now known that this behavior is a natural property of what has been termed
the *DX center,* which is discussed in Chapter 2.

To illustrate the kind of integration of phenomena that is the intent of
this book, we now summarize the observation of a different kind of metastable property in a very different material: the Staebler–Wronski, or optical-

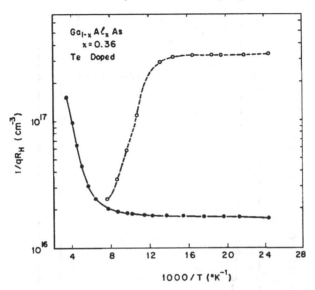

Figure 1.1   The inverse Hall coefficient ($1/qR_H$) as a function of temperature for a
sample of $Al_{0.36}Ga_{0.64}As$:Te. The solid curve represents data taken with the sample in
the dark; the dashed curve is for data taken after the electron density has been saturated by photoexcitation at low temperature. *Source:* Nelson (1977).

degradation effect in hydrogenated amorphous silicon (a-Si:H) (Staebler and
Wronski 1977, 1980). Without the incorporation of hydrogen, a-Si has such a
high density of defects, because of the lack of long-range order, that the material cannot be doped or otherwise used for semiconductor applications. When
hydrogen is incorporated up to proportions as high as 20 percent, the density
of defects is reduced by several orders of magnitude, and the material does
exhibit many of the properties of a typical semiconductor. This material has
shown considerable promise for application in low-cost photovoltaic cells, as
well as in thin-film transistors; but the phenomena involved in optical degradation and current-related defect-forming processes are a major impediment
to its success, as well as a severe challenge to our understanding.

Data reported by Staebler and Wronski on the dark and light conductivities
as functions of the time of exposure to strong light are shown in Figure 1.2.
Both of them decreased markedly, the dark conductivity by several orders of
magnitude. They also reported that these changes could be annealed away by
heating the material to about 150 °C for a short time, and the process could
be repeated; hence, the photoinduced state is metastable. In subsequent years
many other properties were found to change along with the dark conductivity
and the photoconductivity. One key aspect was the observation in kinetics

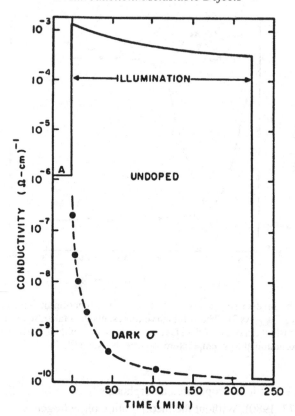

Figure 1.2 Variation of the photoconductivity (solid curve) and the dark conductivity (dashed curve) as a function of time under illumination for $0 \leq t \leq 225$ min for a sample of undoped a-Si:H, illustrating the Staebler–Wronski effect. *Source:* Staebler and Wronski (1980).

studies of nonexponential time dependences for both the onset and decay of the metastable state of the material. Although the agent responsible for these changes in a-Si:H has been identified as a deep-level defect known as the *dangling-bond defect,* the understanding of this case is not as advanced as that for the DX center.

These two selected examples of photoinduced effects exhibit the kinds of phenomena and the types of metastable photoinduced defect properties that this book treats. Many other related phenomena exist, and it is appropriate at least to identify them by name and general characteristics before proceeding to a more detailed analysis of specific results and models.

One way of distinguishing phenomenologically between two major types of effect is by the role of thermal effects in the process of inducing the meta-

stable state of defects. In a variety of effects found in CdS and related materials (see Chapter 3), for example, the formation of the metastable state of a defect depends on both photoexcitation in the appropriate energy range and on
thermal energy (lattice vibrations) available; that is, a temperature threshold
can be defined below which the metastable state does not form (Bube et
al. 1966). For DX defects in GaAs and related materials, a low-temperature
threshold for formation does not occur because photoinduced interactions are
initiated by localized, dramatic multiphonon recombination processes (Henry
and Lang 1977). In a-Si:H, the occurrence of the interaction again depends
only on photoexcitation (or other energetic excitation) and recombination,
and does not exhibit a low-temperature threshold (Benatar et al. 1991). In all
three types of material the effects of the photoinduced interaction can usually
be thermally annealed by raising the semiconductor to a temperature above a
high-temperature threshold for annealing, so that the defect returns from the
photoinduced metastable state to the ground state.

A variety of measurable electronic effects are associated with photoinduced
changes in the properties of defects in a semiconductor. Since the formation
of additional defects often involves a change in the carrier lifetime through
altered capture rates, it is common to find changes in photoconductivity, photoluminescence, or transport phenomena involving carrier collection. The existence of such defects may often cause the Fermi energy to change, with corresponding changes in the measured dark conductivity. Properties of defects
are often revealed in high-resistivity semiconductors through measurements
of thermally stimulated conductivity, and dramatic changes in this conductivity often accompany photochemical defect formation. Photoinduced defects
sometimes have unpaired spins, and hence can be detected through measurements of electron spin resonance (ESR). Related measurements of capacitance, photocapacitance spectroscopy, or deep-level transient spectroscopy
(DLTS) for semiconductor junctions are major sources of information on the
effects of changes in defect charge states.

Several review articles on photoinduced defect mechanisms are available,
some of which are referred to later in more specific topics (Baraff 1986; Lang
1992). In cases like the DX center in III–V compounds, the effects are associated with the properties of impurities that can exist either in substitutional
shallow-level sites or in restructured interstitial sites corresponding to deep
levels, with accompanying large lattice-relaxation effects (Lang 1992). In the
case of a-Si:H, photoexcitation results in the formation of a broken bond with
metastable defect characteristics of its own (Street 1991b, p. 213). In CdS,
typical processes involve local rearrangement of complex defects, association
of previously separated defects, or dissociation of previously associated de-

fects (Sheinkman 1987; Boer 1990, p. 1101). Some of the most pronounced photoinduced defect interactions in II–VI materials occur in materials previously subjected to high-energy electron damage, thus forming metastable defects of types and densities that may not be common in as-prepared material.

1.4 Properties of Localized Defects

In developing the present detailed understanding of PPC and the DX center, all traditional explanations of the behavior of one-electron systems were found inadequate. Instead it became necessary to add another feature to descriptions of deep-level defects: lattice relaxation at localized defects. That is, the local structure in the vicinity of a defect is important, and it can change markedly upon a change in the electronic state of the defect that itself may be induced by light.

A full description of changes in a defect must include changes in both structural and electronic properties. Visible light interacts principally with electrons rather than with the heavier atoms, so the coupling of these two aspects of a defect are of central importance for localized, deep-level defects. For semiconductors, the term *photochemical effect*, which is sometimes used, implies that, along with any possible structural change induced by light, some electronic change occurs, and thus some alteration of the electronic energy levels or their occupation. In traditional treatments, optical effects were thought to change only the occupation of states with fixed energy levels, but now it is necessary to consider possible changes in the allowed energies themselves if neighboring atoms change their positions. This is why one-electron models alone cannot encompass the new effects.

As a practical matter, emphasis is generally placed on defects whose levels lie within the electron-energy gap, since they produce the most pronounced electronic effects. Among the defects formed by foreign atoms there are two broad classes: *dopants,* which have "shallow" electronic energy levels (i.e., the levels lie close to a band), and others whose levels are "deep" in the energy gap. An interesting case is the DX center in III–V compounds, for which the same foreign atom can have either a shallow or deep level, depending on its atomic environment, which in turn can be influenced by any of several factors. We shall see that this possibility bears closely on important properties of some II–VI compounds as well. The aforementioned dangling-bond defect in amorphous silicon can have three charge states, or it can disappear (i.e., become inactive) upon annealing; in this case a "latent defect" remains that can be reactivated by light.

Defects that interact with light in the ways of most interest are generally those whose electronic energy levels are deep enough in the energy gap to permit photoinduced changes to last observably long times. Clearly, this emphasis implies a dependence on temperature, and we shall be concerned with properties of materials generally at temperatures of 20–400 K. We shall see that there are cases for which there is an energy barrier for decay from such a change that is larger than the binding energy of an electron in a donor, so the distinction between shallow and deep must sometimes be qualified.

Typically, deep levels are able to serve as recombination centers for mobile carriers; this is one major evidence of their presence. Shallow levels usually function as traps, as stated earlier. Generally speaking, the defects of interest for photochemical interactions are *not* in the category of traps because their energy levels are normally deeper in the gap, so their electrons are not readily excited into a band by thermal energy. Traps do, however, affect the availability of electrons for capture by deep levels, and it is necessary to account for their presence in analyzing kinetics of electron population changes through recombination centers.

For the sake of completeness, we mention defects in insulators, which can also have significant optical interactions, such as color centers. Because our interest is with semiconducting materials, conventional insulators are not considered, although some semiconductors do exhibit insulator-like properties that are discussed here.

1.5 Shallow and Deep Levels

In discussing various dopants and defects, we shall speak generally in terms of electrons, with the extension to holes always implied. Often a donor atom occupies a substitutional lattice site and has one more electron than the host atom that it replaces at that site. It is of central importance that the main binding force for an electron to a *shallow* donor is the long-range Coulomb force. The electronic states are derived principally from the nearby conduction band; their energies are close to the energies of a Rydberg series, modified from the hydrogen atom in the usual way by the dielectric constant of the material and effective mass of the electron – hence the common term *hydrogenic levels.*

Such donors produce an effective-mass state with a small (shallow) binding energy for electrons that is nearly fixed in the presence of changes in temperature or hydrostatic pressure or alloying. A major feature of hydrogenic states in semiconductors is that their radii are large compared to the atomic spacing; only in that way does the dielectric constant of the macroscopic material become effective. Therefore, shallow levels and large orbits go together.

One consequence of the large orbits is that motions of the atoms of the host material have slight effects on the behavior of these electrons.

There are no changes in the atomic positions around shallow donors resulting from optical interaction: The position and coordination of a substitutional donor atom remain unaffected when light changes the large-orbit electronic state. At temperatures high enough for rapid thermal exchange with the adjacent band, optically induced changes in electronic occupation are unimportant since they are promptly overwhelmed by thermal effects. Hence low temperatures are required for observation of the optical properties of shallow donors. In most cases it is clear whether or not thermal changes need to be considered, although current understanding tells us that an exception must be made for the DX center in III–V compounds.

Deep levels differ from these in essentially all ways. Their states are composed of contributions from both valence and conduction bands, and perhaps from other bands as well. The dominant binding force for an electron is not Coulombic, but short-range; thus the lowest levels do not correspond to a hydrogenic series. Since the binding force is not Coulombic, the charge state of a defect is not necessarily the determining property for its capture of charge carriers. There is considerable confusion in the field over the magnitudes of capture cross sections, in part because of this property. Typically the radii of deep-level states are near atomic sizes; that is, these states are *localized*. This localization is a more important property than the energy level of the defect state, and there are cases of such localized states having energies that are resonant with a crystal band; then terminology becomes a problem because some workers use the term *deep* to mean localized.

The localized character of the electronic state of a deep-level defect makes it much more sensitive to the positions of the neighboring atoms; in this respect these defects have properties like those of ionic materials, and the interactions of electronic and configurational changes (i.e., electron–phonon coupling) are of paramount importance. This importance had not been widely recognized before the elucidation of the DX center, but it has become the dominant new feature in explaining metastable effects. The higher excited electronic states of deep-level defects, however, may have larger orbits than that of the ground state, and for them electron-lattice effects become less important, whereas Coulombic effects are larger. An older theory for the capture of a free carrier by a Coulomb-attractive localized center held that it takes place in stages, with the initial capture into an excited large-orbit state, followed by sequential relaxations into the ground state by successive emissions (a *cascade*) of phonons (Lax 1960). However, more recent work has established that capture most often occurs by multiphonon emission, as described here. We give an introductory discussion that is based on those of Lang (1992) and Baraff (1986), who have done some of the key work in developing these ideas, and on a review by Langer (1980).

Although these relations and comparisons are phrased in terms of crystalline semiconductors, in good amorphous semiconductors the similar bonding and energy bands make these same principles applicable. In the important case of a-Si:H, both doping and deep levels occur in very much the same way as for crystals, although just one type of deep-level defect (the dangling bond) appears to dominate its properties. The photoelectronic properties of these deep-level defects in a-Si:H are of vital importance because of their impact on major photoelectronic applications that are under development.

For all of these reasons, the defects of interest in this book generally have deep levels and are metastable.

1.6 Metastable Defects and Configuration-Coordinate Diagrams

Optical interactions can have major effects on the structural properties of deep-level defects, in contrast to shallow ones. When light causes an electronic excitation of a localized defect, the alteration of the electron wave function induces changes in the surrounding distribution of electronic charge, which cause movement of the neighboring atoms, that is, a *configuration* change. This process is generally referred to in abbreviated form as the *electron–lattice interaction* at the defect. Typically it is local modes of vibration that are most important for this electron–phonon coupling rather than the crystal modes. If, for example, the defect has symmetrically located neighbors, an excitation may alter the nearest-neighbor distance symmetrically, producing a *breathing mode* of configuration change. This happens in some insulators, but not for the vacancy in Si, where the degeneracy of the tetrahedral bonds causes the symmetry to drop around a vacancy by the Jahn–Teller effect (i.e., any system having a degenerate ground state will spontaneously distort and lower its symmetry). An essential point is that the term *configuration* of a defect refers to its position and the positions of all the atoms in its immediate vicinity. If, for example, a hydrostatic stress is applied to the material and all the atoms move closer, that represents a new configuration even if it retains all the initial symmetry (which is not always the case). For these reasons, we shall refer to the *defect center,* which includes the central defect itself and its immediate neighborhood.

A new configuration induced by an electronic excitation may or may not be relatively stable; if the excited electronic state with the new configuration can promptly (relative to the observation time) relax back to the ground state, it is not stable. The cases of interest here are those for which a new configuration is metastable, so that it will endure long enough to be observable. In these instances, understanding a defect requires more than answering the questions What is it? and Where it is located? Just as important is, How does it work? That is, what are the rules that control transitions between its ground and metastable states? Since these transitions are accompanied by observable changes in material properties (as illustrated in Figs. 1.1 and 1.2), they provide a prime source of information on metastable defects, and kinetics data are often the basis for later identification of the nature of the defects. A major example of this is the DX center in III–V compounds: The transitions of initial interest involve kinetic relations governing capture or release of electrons, and accurate measurements of these are the key to identifying the multiphonon emission mechanism as the controlling one for changes in state (Henry and Lang 1977).

Metastability implies that there is some property that inhibits relaxation to the ground state. In atomic transitions, such an inhibition arises from selection rules that forbid a transition, and similar suggestions have been advanced for defects in solids; but nearly all recent work in semiconductors shows that defect metastability generally implies the existence of an energy barrier between the ground and excited electronic states that must be overcome for relaxation to occur. Moreover, this barrier is associated with a *large lattice relaxation* (LLR) of the defect center that occurs during an electronic change of state (Henry and Lang 1977; Mooney 1990; Bube 1992, pp. 175, 270; Lang 1992). All of this is represented simply in a configuration-coordinate (CC) diagram like Figure 1.3 for a single center. In this context the term *relaxation* refers to a change in energy and configuration of the center that is associated with a return to the electronic ground state of the defect.

In CC diagrams, the ordinate is the Gibbs free energy of the solid, but entropic and volumetric changes are always ignored as being small, so it is the total energy that is plotted. This total energy explicitly includes both the electronic and strain energies of the center. Although the abscissa is called a configuration coordinate Q, it is not an ordinary Cartesian coordinate. Rather, it is a symbolic representation in one dimension of the entire set of real coordinates that specify the positions of the defect itself and all the nearby atoms that interact with it. Hence a value of Q symbolizes a complete configuration for those atoms, and any change in Q implies a possibly complicated set of changes in atomic positions. The simplest example of this idea is a symmetrical defect in which all the nearest neighbors are equidistant from the central site. Then Q might represent the nearest-neighbor distance, and its change would describe a breathing-mode change in the atom positions around the

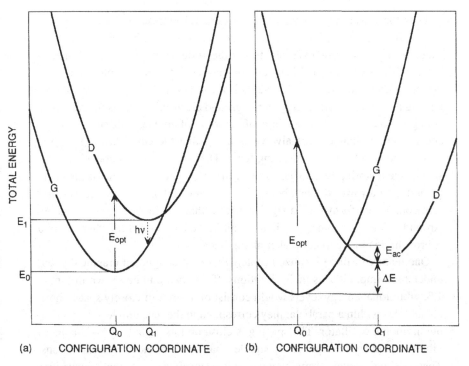

Figure 1.3 Configuration-coordinate diagrams of a localized center having (a) small or (b) large lattice relaxation. E_{opt} is the minimum excitation energy out of the ground state by a photon, and ΔE is the minimum thermal ionization energy.

center. Generally, the situation is much more complicated, and each value of Q must be thought of as symbolizing a three-dimensional snapshot of the group of atoms.

A major property of CC diagrams is their assumption of the Born–Oppenheimer (adiabatic) principle that electronic energies can be evaluated as if the atoms are fixed in position, with any change in configuration permitting a reevaluation of the electronic energies in the same way, but using the new positions. This explicitly ignores contributions to the energy from the kinetic energy of the atoms. One significant case in which this assumption had to be dropped, mentioned in Chapter 2, is the detailed quantitative characterization of the DX center. We shall deal with these diagrams at some length because they are less familiar in discussions of semiconductors than for insulators and because there seems to be no thorough introductory treatment that is designed for semiconductors. It is also necessary to establish the link between the CC descriptions of localized defects and energy bands.

Thus the lower curve labeled G in Figure 1.3 represents the total system energy for the state of the solid in which the electrons are in their ground state. To take a specific example, this is the system energy when the valence band is full, the conduction band empty, and the defect unoccupied. The configuration in which the atoms have their ideal thermal-equilibrium positions, leading to the minimum energy of the ground electronic state, is the reference configuration Q_0. Any change in configuration, of course, adds strain energy, and for small changes it is always assumed that the curve has a parabolic shape produced by a fixed restoring force. The width of the parabola is a measure of the restoring force: The stronger the force, the sharper the rise in energy and the narrower the parabola. The normal covalent bonds of tetrahedral semiconductors form such a rigid structure that the force constant is large. Around a defect, this may be altered, and in the case of a-Si:H where a bond is broken, it is likely to be weaker along that axis.

One feature of these localized centers that is often ignored graphically, but understood implicitly, is their vibrations. If they were to be shown in Figure 1.3, vibrational energy levels would consist of a series of closely spaced horizontal lines within a parabola; they correspond to the quantum levels of a simple harmonic oscillator. The spacing is close on this scale because quanta of vibrational energy are generally smaller than those of electronic excitations. These vibronic states permit excursions in Q away from Q_0; these excursions are limited classically by the line of the parabola. Because of their presence, a photon with energy higher than the minimum shown in Figure 1.3 can be absorbed into a final state above a parabola having a higher vibrational state of excitation.

The upper curve labeled D in Figure 1.3 represents the total system energy for the first excited electronic state; for the example here, we take it to have one electron localized in the defect center and one valence-band hole that is mobile and not interacting with the defect (ignoring the state in which hole interaction does occur). The energy minimum occurs at a configuration Q_1 different from that of the ground state for reasons already discussed; but electronic transitions must be vertical in these diagrams (the Franck–Condon principle) because atoms are effectively stationary during the short time of an electronic transition. Hence optical excitation as shown requires considerably more energy E_{opt} than the difference between the minima ΔE, which is the *thermal* ionization energy. (There are cases for which this optical ionization energy is even larger than the band gap of the semiconductor.) Once optical excitation occurs at the original configuration Q_0, the preferred configuration becomes Q_1, so the defect is in a highly excited *vibrational* state. Hence there is a rapid decay in vibrational energy by phonon emission as the center seeks

Q_1, E_1. These phonons are large-amplitude, predominantly local modes associated with the localized center, and their effect can be thought of as violent rattling of the atoms in the vicinity. This decay process is called *multiphonon emission* and is central to the initial explanation of the properties of what is now called the DX center (Henry and Lang 1977).

After relaxation to Q_1, E_1, there are two possibilities for decay out of the photoexcited state: radiative and nonradiative. To treat them we first must distinguish between cases for small and large lattice relaxations shown in Figure 1.3. In the small-relaxation case, Figure 1.3(a), the minimum of curve D lies within the parabola of G so that there is overlap of the two sets of vibronic states near Q_1. Here decay from the excited state D can occur by either of two competing processes:

1. nonradiatively by transitions from D to G permitted at any temperature by the overlap, followed by phonon emission to the bottom of G; or
2. by luminescent emission of a photon of energy $h\nu$ as shown, at constant Q_1, followed by phonon emission.

This photon always has lower energy than that for optical excitation – a relation called the *Stokes shift* ($E_{opt} - h\nu$). If the initial excitation is to an energy on D that is above the crossing of G and D (as in the case shown), there will be little luminescence (Stoneham 1985).

For the case of large lattice relaxation shown in Figure 1.3(b) the minimum of D lies outside the parabola of G, so there can be no radiative decay from D to G at the configuration Q_1. Since there is no overlap of vibronic states of D and G near Q_1, the only path for decay is the nonradiative one that requires thermal energy in the amount E_{ac} as shown. It is clear that in this way E_{ac} becomes an energy barrier for thermally activated relaxation to occur from the excited electronic state, so that once the defect is in this state, it will persist if thermal energy is low; this is the metastable condition that is central to the photoinduced effects of interest in this book. (The term *bistable* is sometimes used instead of metastable for such a center.) Such metastability is the accepted explanation for persistent photoconductivity in homogeneous materials.

1.7 Energy Bands and Configuration-Coordinate Diagrams

For insulators, CC diagrams of defects deal principally with the internal workings of the defect, but in semiconductors it is essential to relate these also to the energy bands of the host material. Because this is rarely done in books, we have enlarged this discussion to provide an introduction to this relationship. One facet of this preparation was the identification of the ground state in

Figure 1.3 as having a full valence band as well as an unoccupied defect. (We note in passing that to be sure that any defect is unoccupied, the Fermi energy is taken to be at the bottom of the energy gap; we shall return to this implication in Section 1.8.) The next step is to specify that the energy difference ΔE in Figure 1.3 excludes any kinetic energy of electrons within the valence band, so ΔE is the *minimum* that can form the excited state. This implies that the electron being excited has its initial state at the top of the valence band, and hence that the curve labeled G in Figure 1.3 may be taken to be the energy of the top of the valence band E_V. This is equivalent to choosing a reference energy at the top of the valence band in equilibrium.

The minimum conduction-band energy E_C of a semiconductor represents the state in which the conduction band has a single electron, the valence band has one hole, and the carriers are far apart so as not to interact. Because these carriers are not localized, this excitation has negligible effect on the configuration of a center, in contrast to the promotion of an electron to a localized state. Hence E_C can be represented by another parabola identical to that for E_V since it represents just the local strain energy, centered at the same equilibrium configuration but higher in energy than E_V by just the value of the energy gap, as shown in Figure 1.4. To complete the description of the state with energy E_C, we must also specify the occupation of the defect, which is unoccupied because only one electron has been excited from the ground state.

Similarly, a full description of the defect state D includes a specification of the occupation of both bands when the system is in state D. In the present context, excitation of a single electron from the valence band into the localized center to form D implies that the conduction band should be empty. Thus we have also related the configurations to the known property that a deep level typically provides a state having lower energy than an interband transition.

In the same way that excitations from the valence band to the defect state are described by these diagrams, we can extend the description to excitations from the defect level to the conduction band. It must always be borne in mind that any electronic transitions to or from a deep level may involve a significant lattice relaxation of the neighborhood. We shall see later that the consequences of such relaxations can be profound, explaining, among other things, persistent photoconductivity and the temperature dependences of capture cross sections of carriers in many cases, as well as providing the only explanation of negative effective electron correlation energy for a two-electron center.

It is worth emphasizing that the energy represented by a CC curve at *any* configuration is the energy of a quantum state of the material (including the

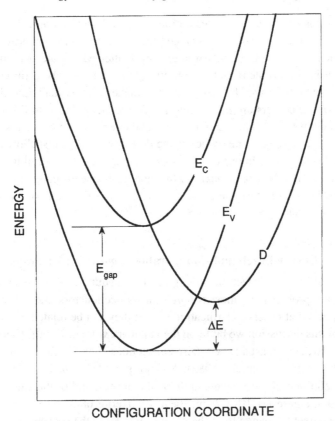

Figure 1.4 Configuration-coordinate diagram of a localized defect with a large lattice relaxation, in relation to the nearest band edges E_V and E_C.

central atom and its neighbors) for specific values of electronic excitation and atomic positions. It may happen that the center exhibits observable properties for only some of these electronic states and not for others. If we hypothesize that the defect of Figure 1.4 is electronically active when the system is in state D (i.e., when an electron occupies the center) but inactive in the ground state G, then we would be unaware of the presence of a *defect* until some excitation occurs. In such a case it would appear that a defect is *created* by the excitation even though a departure from perfect structure is present in both conditions. This happens in a-Si:H, for which the ground state of the principal center is normally inactive (unobservable) and the photoinduced state of the center is active.

Terminology here is sometimes a problem; is a defect *created*, to use the common term (Adler 1984)? Our choice here is to use the broader term *center* for the departure from perfect structure, no matter what state it is in, and whether it is observable or not. The term *defect* is then reserved for centers that are in a state in which their properties make them evident, and the remainder of the centers are *latent defects*. To reconcile this with DX-center terminology, we note that by common usage the term *DX center* refers to the center in its deep-level state, which is the defect, whereas the shallow-donor state is a latent defect. In our terminology, we might choose to call it the *DX defect* and reserve the word *center* for the more inclusive meaning of either of the states, but in this case the present terminology is too well entrenched to propose such a change.

1.8 Energy Levels and Configuration-Coordinate Diagrams

Although diagrams of electronic energy levels in equilibrium as functions of position are probably the most common in semiconductors, there are some basic aspects that deserve clarification so that they can be related to CC diagrams. In this discussion we build on the landmark work of Baraff, Kane, and Schluter (BKS) who established these connections (Baraff, Kane, and Schluter 1980; Baraff and Schluter 1985). A major point to be made is simply the correct definition of these levels in thermal equilibrium. For the simple case described in Figure 1.4, BKS define the *level* in terms of the energy needed to cause the defect to change from the ground state E_V to the next-higher-charge state D; hence it is simply ΔE, the thermal ionization energy of the center, measured from the top of the valence band. Thus with Figures 1.3 and 1.4 we see three kinds of energy differences that must be distinguished when lattice relaxation is significant: the optical excitation energy E_{opt}, the activation energy E_{ac} (a barrier height), and the ionization energy ΔE. It is only in the framework of a CC diagram that their relationships can be fully clarified.

In equilibrium how can state D become occupied; that is, where can this energy ΔE come from? There are two possibilities (since equilibrium excludes external excitation of any sort): thermal energy at finite temperatures, or at low temperatures, doping. It is this low-temperature, doping-controlled behavior that BKS treat, and ordinary temperatures do not alter the relations in principle. Since n-type doping raises the Fermi energy E_F so that eventually D must become occupied, it is E_F that provides the extra energy ΔE in Figure 1.4 because the chemical potential of electrons in the system increases.

This is the key concept that BKS included in calculations of the total system energy as a function of doping. When E_F is low in the gap, the occupation

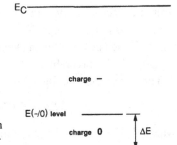

Figure 1.5 Energy-level diagram for a center with two possible states having the charges shown, depending on the value of E_F.

probability of a gap state is low, and the unoccupied center is in the electronic state that has a neutral charge (if this is an acceptor-like state). At some higher value of E_F the occupation probability (at low temperature) switches from zero to one. The "level" is defined as *the energy that E_F must have for the electronic state to change*. Because this change alters the charge on the center from neutral to negative, the energy of the level is denoted by $E(-/0)$. This is then the designation for an acceptor-like defect level at ΔE above E_V as illustrated in Figure 1.5. An essential principle is that *all levels should be identified with their two relevant charge states, not a single charge*. A level does not describe a single state of a center, but rather the energy that E_F must have for the switch between two charge states. It is improper to refer to a level by a single-charge designation; such a designation, such as D^0 or D^-, applies to the charge *state* that exists for a range of values of E_F. We thus distinguish henceforth between the charge state of a defect and the energy levels that characterize transitions from one state to another.

Although it is obvious that E_F controls the charge of the center by controlling the electron occupancy, this definition of a level may seem somewhat unfamiliar, and it is instructive to consider the underlying reasons that E_F is central to the definition of a level. One approach to this question is by comparing an optically induced change-of-state transition of the type discussed in connection with Figure 1.3 to the thermal-equilibrium change of state that occurs as the result of a doping-induced increase in E_F (at low temperature). In the optical case a quantum transition occurs from a definable initial state E_V to the defect state. In the equilibrium case, however, something entirely different happens, aside from the reconfiguration that has time to occur. An electron that goes into the defect must come from the entire system of electrons that serves as a *reservoir* in the thermodynamic sense. The incremental energy change of any such reservoir upon the loss of a single particle is the chemical potential of that species, which for electrons in a semiconductor we denote by E_F; thus an electron captured at an isolated center withdraws an energy E_F

from the reservoir. The total energy change of the defect plus reservoir in this transition is $[E(-/0) - E_F]$. If the localized state has energy ΔE, then the center will become occupied by an electron only when $E_F > \Delta E$ so that the transition is energetically favorable.

This is expressed in the relation for the total energy of a state,

$$E_{tot} = T + V + \Phi + V_I + E_{xc} - N E_F \qquad (1.1)$$

where T, V, Φ, V_I, and E_{xc} are the electronic kinetic, electron–ion, electron–electron, ion–ion, and exchange-correlation energies, respectively; and N is the number of electrons in the center ($N = 0$ or 1 here, and 0, 1, or 2 for a center that can have three states of charge). Minimization of the total energy of the combined system of defect and reservoir must include this last term. Thus *any variation in E_F causes a change in the energy of each state that exchanges electrons.* When E_F lies precisely at a *level*, the empty and occupied states that define that level have equal energies. When E_F moves above a level, minimization of the system energy requires that an electron transfers from the reservoir into the defect and thus changes its state. Notice that *the level does not shift with E_F*; instead, the shifting energies of the states define the level when two of the energies are equal. BKS presented a graphical way of representing these relationships (Baraff, Kane, and Schluter 1980).

It may seem that changes in E_F might cause corresponding changes in optical energies as the energies of the states vary, but that is not so. All these considerations are necessarily equilibrium relations, since E_F would not otherwise be defined. To deal with nonequilibrium processes like optical excitation and DLTS activation energies, one must exclude E_F from the calculations of Eq. (1.1); the resulting energetic relations then form a true configuration-coordinate diagram from which barrier heights and optical excitation energies can be inferred. This is done by setting $E_F = 0$, which sets it at the top of the valence band, the energy reference.

The vacancy in crystalline Si as described by BKS is a center that has three possible charge states. We now mention another center that is either negative, neutral, or positive, depending on whether the number of electrons occupying it is two, one, or none as illustrated schematically in Figure 1.6. This describes the dangling-bond defect in a-Si:H, which is discussed in detail in Chapters 4 and 5; the (inactive) ground state of the center is additional. The three charge states of this example imply two levels for this defect, designated $E(0/+)$ for the lower one and $E(-/0)$ for the upper one. Each level is the thermal energy required to add an electron in equilibrium to the defect from a reservoir whose chemical potential is coincident with the top of the valence

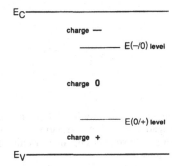

Figure 1.6   Energy-level diagram for a center with three possible states having the charges shown.

band. When E_F varies, what changes is only the relative ordering of the energies of the three states: when E_F is in the middle range, for example, the states with negative or positive charge are not totally absent; they simply have higher energies than the neutral state (see Figure 4.3). Similarly, when E_F is in the lower range (i.e., below the $E(0/+)$ level) the positively charged state acquires a lower energy than that of the neutral state (or the negative state). These relations are discussed further for a-Si:H in Section 4.2.

Baraff, Kane, and Schluter (1980) give detailed descriptions of further consequences of these relations. Specifically, "*this definition* [of a level] *includes both electronic and lattice readjustment energies and thus is not given by a one-electron eigenvalue calculation*. It is *only* these levels (characteristic of two charge states), and *not* the one-electron eigenvalues, which can be measured in a quasi-equilibrium experiment" (italics in original). Also emphasized by BKS is the distinction between these levels and observed activation energies (e.g., for transient capacitance changes). The activation energy is the slope of a logarithmic change of some rate process when plotted against reciprocal temperature. In conventional semiconductor analysis this distinction is ignored, but for deep levels in which there may be an energy barrier to thermal excitation (or relaxation) of electrons, as in E_{ac} in Figure 1.3, it can be essential. From the often-used DLTS, an activation energy is obtained (corresponding to E_{ac} in Fig. 1.3), not an equilibrium energy difference.

There are defects, however, for which the simple ordering +, 0, − of the charge states shown in Figure 1.6 is not followed; these have a *negative effective correlation energy* (*negative U*). In this situation the lattice relaxation induced by changes in charge state is so important that it becomes more favorable energetically for the center to have two electrons than to have one, despite their repulsive electron–electron correlation energy. Figure 1.7 shows a level diagram for a negative-*U* center that can have stable states that are either neutral or doubly negative; here a singly negative state must be at some

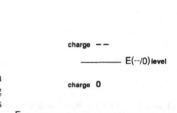

Figure 1.7 Energy-level diagram for a center with negative effective correlation energy, thus having an equilibrium occupancy of either two electrons or none.

still-higher energy. The negative-U interpretation by BKS for the vacancy in Si is now widely accepted, and it appears to be common for defects in chalco-genide glasses as well.

One further facet of this kind of analysis deserves mention. In Baraff, Kane, and Schluter (1980) implicitly, and in Baraff and Schluter (1985) explicit-ly, entropic considerations were omitted in calculations of the Gibbs free en-ergy that should be used as the ordinate for these diagrams. Baraff and Schlu-ter (1985) state that "uncertainties in the calculated values of [the energies] already exceed any reasonable estimate of $T\Delta S$," where ΔS is the entropy change. Also, their calculations are for the limit of low temperatures. Since electronic energies of interest for deep levels in semiconductors are typically ten times the value of kT at room temperature, these approximations seem ap-propriate.

Some of this description of levels applies also to shallow (hydrogenic) lev-els. Specifically, the definition in terms of the value of E_F is the same. An im-portant difference is that changes in the electronic state of shallow levels are not associated with local configuration changes because of the large orbits of electrons in shallow states. Thus, a shallow donor level would be represented in a CC diagram like Figure 1.4 by a parabola slightly below E_C having the same shape and no displacement in configuration space. In an energy-level diagram both shallow and deep levels are represented in the same way, and the possible configurational effects are not visible. This makes the ordinary level diagrams less useful when relaxation effects are significant. An interest-ing complication occurs for the donor-related DX center in GaAs upon either alloying with AlAs or application of hydrostatic pressure. Then changes in the local bonding of the donor atom can cause a substantial change in its con-figuration, which produces a shift of the lowest electronic state of the center from that of a shallow level to one with a deep level. This change (described in detail in Chapter 2) accounts for essentially all of the major properties of the DX center.

Table 1.1. *Examples of defects*

Specific defect	Ground state G	Metastable state M
DX in GaAs	Donor in substitutional position	Donor in interstitial position
DX in $Al_xGa_{1-x}As$ with $x > 0.2$	Donor in interstitial position	Donor in substitutional position
DX in ZnSe	Acceptor in interstitial position	Acceptor in substitutional position
Dangling bond in a-Si:H	"Weak Si bond"	Dangling Si bond
Defect in CdS:Cu	(One associated defect complex)[a]	(Two dissociated defects forming an electron and hole trap)[a]

[a]A possible illustrative model.

1.9 Density of States

In situations for which lattice relaxation is significant for transitions between states, the identification of a density of states for optically induced transitions becomes complicated. This quantity $N(E)$ describes the number of states per unit volume per unit energy range for initial and final states participating in a transition. In general, the transition is likely to involve changes in both electronic and vibronic states. The density of states in any given energy interval depends on the placements of the electronic parabolas as well as on their shapes, which reflect the density of vibronic states. These shapes are rarely used, or even known. If a distribution of configurations is possible among centers of interest, the shapes may have a corresponding distribution. Also, where (the very likely) departures from idealized parabolas occur, the density of vibronic states varies with energy as well. Therefore the strength of an optical transition may not be a simple measure of the defect density.

1.10 Summary

To summarize the concepts involved in the variety of metastable defects described in this book, Table 1.1 presents some typical characteristics, the details of which are given in later chapters.

2
III–V Compounds: DX and EL2 Centers

2.1 Introduction

Among defects in semiconductors the DX center is unique in the extent and detail of its studies and in the advanced present state of its understanding. It (or similar centers) has been observed in n-type III–V compounds doped with a variety of species of atoms and in some II–VI compounds; it may also be related to the dominant defect in a-Si:H. The donor-related DX center in AlGaAs alloys (the most widely studied) can thus serve as a model that can usefully be invoked in the study of other semiconductor defects; the EL2 defect in GaAs – a native defect in undoped material now thought to be an As_{Ga} antisite defect (an As atom on a site that normally would have a Ga atom) – has some similar properties, and is discussed in Section 2.6. We show later in this chapter that this model provides a new and convincing explanation for a major limitation on the ability to dope some III–V materials. This limitation has caused even more serious impediments to the applications of II–VI materials, and (as discussed in Chapter 3) the DX model is useful there too. Present understandings followed nearly twenty-five years of efforts to elucidate metastability in III–V compounds, and the literature is large; we therefore present here only a brief summary of the early work, which is described well in reviews by Lang (1992), Langer (1980), and Mooney (1990, 1991), who also discussed the impacts of such deep donors on the properties of devices using these materials.

A number of observations of anomalous effects have been reported for these centers, with the consequent conclusion that the familiar one-electron picture of deep-level centers is inadequate. The key anomalous effect is metastability (occasionally called *memory effect*) in several properties, the most dramatic of which is persistent photoconductivity (PPC). This was reported in CdS by Litton and Reynolds (1962) and nonstoichiometric CdTe by Lorenz,

24

Segall, and Woodbury (1964), and was illustrated in Figure 1.1 for AlGaAs alloys from the work of Nelson (1977). This increase in conductivity caused by subgap light can last for exceedingly long times at low temperatures; thus the term *bistable* is sometimes used for the material at these temperatures.

Craford and coworkers (1968) reported the first experiments on what would later be called the DX center, describing PPC in n-type GaAs$_{1-x}$P$_x$ alloys at temperatures below ≈ 100 K. That paper showed that major parameters affecting these centers are alloying and pressure, parameters that have since been used to clarify considerably the nature of these centers. In addition, in the work from which Figure 1.1 was taken, Nelson reported that the electron mobility was considerably higher in the dark at low temperature, by an amount that is quantitatively consistent with a change in impurity scattering caused by a density of scatterers comparable to the change in free-electron density. This identified the center responsible for PPC as a deep donor. The nonequilibrium distribution of electrons represented by PPC could be induced by either optical excitation or rapid thermal quenching in the dark (Nelson 1977). Since DX centers were then unknown, Craford et al. (1968) suggested two possible interpretations (which have since been superseded) of the persistence:

1. A Coulomb barrier keeps optically freed electrons away from substitutional donors at which they could be recaptured.
2. Band-structure effects forbid the recapture transition.

In subsequent years, a variety of such purely electronic models were proposed to explain the persistence, that is, the inability of an electron to return to its ground state after it makes a transition. Among the reasons given in early papers for the slow return were symmetry selection rules that reduced the matrix element for the return transition, density-of-state factors, slow intervalley transfer of electrons (Iseler et al. 1972), and macroscopic internal Coulomb barriers due to electric fields between high- and low-resistivity regions of inhomogeneous material (Queisser and Theodorou 1979). All of these proposals have been tested experimentally over the years and been found unsatisfactory for homogeneous material of steadily improving quality. Instead, it is now generally accepted that most metastability in general, and persistence of photoconductivity in particular, is due to an energy barrier associated with the kind of large lattice-relaxation effects that were described by configuration-coordinate diagram in Figure 1.3(b). This picture is supported by many types of observations.

Another property of Al$_x$Ga$_{1-x}$As alloys that earlier seemed unrelated to the metastability is the dependence of doping efficiency on the value of x. When

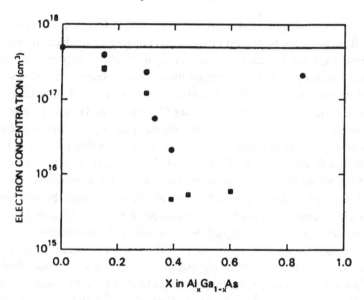

Figure 2.1 The room-temperature electron density in $Al_xGa_{1-x}As$ as a function of the Al mole fraction x in the layer. The solid line indicates the Si concentration measured by SIMS. The data points correspond to the electron density measured by capacitance and Hall effect. *Source:* Kuech, Veuhoff, and Meyerson (1984).

measured by the Hall effect in Si-doped material at 300 K, the electron density is equal to the Si density (determined by SIMS) for $x=0$ or $x=1$, but not in the midrange of composition (Kuech, Veuhoff, and Meyerson 1984), as shown in Figure 2.1. This doping efficiency can be several orders of magnitude below unity and create serious difficulties in the control of conductivity of the material, and in devices that depend on it. We shall see that this is due to the fact that the ordinary donors have *deep* levels, not shallow, in this range of compositions.

There are several complications in the interpretations of all of these observations, and they undoubtedly contributed to the long time that passed until a general explanation became accepted:

1. the previously unknown existence of both shallow- and deep-level states of a single donor, which was the last property to be recognized;
2. band-structure effects, in which the character of donor states might change if they are associated with the X or L minimum of the conduction band rather that the usual Γ minimum;
3. the fact that hydrostatic pressure, which is used to study band-structure effects, also changes the properties of the DX center;

4. the nonexponential kinetics of transitions between the two states of the metastable DX centers, which were difficult to interpret and which broadened DLTS peaks;
5. very different values of some activation energies depending on whether they were obtained from equilibrium or transient measurements (DLTS is a transient measurement), or from thermally or optically stimulated changes; and
6. alloy effects, which influence both the band structure and DX properties ($Al_xGa_{1-x}As$ changes from direct- to indirect-gap for $x > 0.37$, although the DX defect becomes apparent when $x > 0.22$).

It is now known that sufficient alloying or pressure causes the ground state of the donor to change from a shallow-level, effective-mass state to a deep-level, localized state. In discussing these developments here we only summarize the early work (before about 1980), which is well covered in reviews (Langer 1980; Mooney 1990, 1991; Lang 1992); we then show in more detail the ways in which various observations later contributed to the present comprehensive picture.

2.2 Large Lattice Relaxation and the DX Center

The first major step in elucidating the DX center was the demonstration that large lattice relaxation (LLR) is the proper framework within which to describe it; this was done by Henry and Lang in their measurements and analysis of the magnitudes ($<10^{-20}$–10^{-14} cm^2) and temperature dependences of capture cross sections for carriers in n-type GaAs and GaP, and their interpretation that most (although not all) cross sections represent multiphonon emission processes (Henry and Lang 1977).

In these studies the principal experimental techniques were photocapacitance and deep-level transient spectroscopy (DLTS), which permitted evaluation of the densities of deep-level centers and their capture and emission properties, although the nature of these centers could not be identified. Their collected data on the temperature dependence of capture cross sections, presented in Figure 2.2, were analyzed and fitted with a comprehensive LLR model. Henry and Lang emphasized a general principle that had not been widely recognized: "Deep impurities are . . . likely to have substantial relaxation of the lattice equilibrium position near the impurity after capture. . . ." It had been long known that lattice-relaxation effects play important roles in strongly ionic materials, but it was generally thought that in covalent materials the electron–lattice interaction is too weak for such effects to be impor-

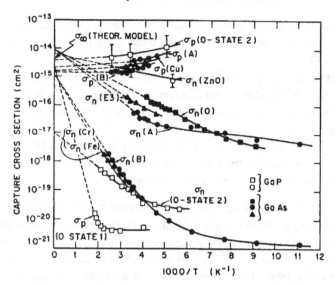

Figure 2.2 Carrier-capture cross sections as a function of inverse temperature for various deep levels in GaAs and GaP. An n subscript denotes electron capture whereas a p denotes hole capture. Values are shown corresponding to Cu, Fe, Cr, O, and the E3 radiation damage defect in GaAs, and also for two unidentified but commonly occurring levels (A) and (B). For GaP the cross sections are associated with the Zn–O center and the two states of the oxygen center. The dashed lines show the temperature dependence of the cross sections extrapolated to infinite temperature, corresponding to the value of σ_∞. Source: Henry and Lang (1977).

tant. However, for reasons given in Chapter 1, spatially localized defect states can indeed have strong lattice-relaxation effects even in these materials. One of the theoretical results of Henry and Lang's analysis was the conclusion that the usual adiabatic (Born–Oppenheimer) approximation breaks down at crossings of configuration-coordinate curves. They developed an alternative treatment for these cases based on tunneling between vibronic states of the defect where anticrossing repulsion occurs.

Henry and Lang did not discuss PPC in that paper; but they soon realized that it was linked to their capture cross sections, and Lang and coworkers proposed an explanation for PPC based on LLR in a deep-level defect that they called a DX center (Lang and Logan 1977; Lang, Logan, and Jaros 1979). The "D" in this name was used because of the clear dependence of these effects on the density of donors, and they proposed that the center is a complex of a donor and some unknown object (X), probably a vacancy. For Te-doped AlGaAs they found that the alloy composition has a dramatic effect on the magnitude of the DLTS signal, as shown in Figure 2.3. In addition, for $Al_{0.37}Ga_{0.63}As$ the photoionization threshold (0.75 eV) was found to be much

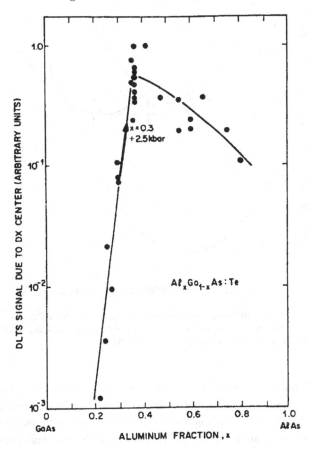

Figure 2.3 The DLTS signal magnitude due to DX centers in various $Al_x Ga_{1-x} As$ samples as a function of the Al fraction x. The heavy arrow indicates the signal increase in an $x = 0.3$ sample induced by the application of a 2.5-kbar stress. *Source:* Lang, Logan, and Jaros (1979).

larger than the thermal binding energy (0.1 eV), and theoretical fits of the shape of the photoionization threshold support the LLR picture that they developed. A number of different dopant species were used with results that were qualitatively similar, so they excluded any chemical properties and suggested that this LLR mechanism explains metastable effects in other materials. Such effects have been reported in CdTe (Lorenz et al. 1964) and in $Zn_x Cd_{1-x} Se$ (Dissanayake et al. 1992). One feature that took much longer to explain is the "markedly nonexponential time dependence" of thermal capture and emission processes of electrons at the center (Lang and Logan 1977).

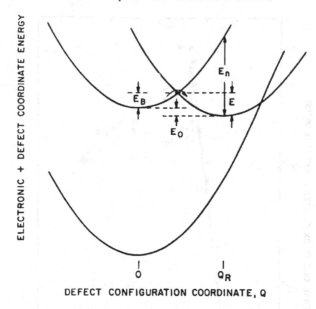

Figure 2.4 Proposed configuration-coordinate diagram for DX centers showing the notation used in this chapter and in Table 2.1. *Sources:* Henry and Lang (1977), Lang and Logan (1977), and Lang, Logan, and Jaros (1979).

   The LLR model proposed by Lang and coworkers for the properties of the DX center in AlGaAs and GaAsP is shown by the configuration-coordinate diagram in Figure 2.4, which applies to conditions in which the deep-level state is the ground state of the center. This occurs with either high pressure on GaAs or alloy compositions of $Al_xGa_{1-x}As$ with $x > 0.22$. This is a specific case of LLR of the type already shown in Figure 1.3(b), and it has an energy barrier separating the minima. The heights of the thermal barriers for electron emission from, or capture by, the center are E or E_B in Figure 2.4. The photo-ionization threshold of the center E_n, can be much larger than the thermal binding energy of the state E_0, which was determined from the temperature dependence of the Hall coefficient. Values of E_B and E were determined from the temperature dependence of the electron capture and emission properties of the center (DLTS). This picture offers a consistency check since $E = E_B + E_0$, and very good agreement with this relation was found for AlGaAs and for GaAsP doped with various donor species, as shown in Table 2.1 (Lang 1992). The photoionization thresholds E_n for different donors are quite different, making chemical effects easy to identify in this way.
   Comparison of E_n with E_0 also permits the inference of the shapes of the parabolas forming the CC diagrams. Even the shape of the onset of photo-

Table 2.1. *Values (eV) of the parameters in the configuration-coordinate diagram of Figure 2.4 for AlGaAs and GaAsP with different dopants*

Donor	E	E_B	E_0	E_n
AlGaAs				
Se	0.28 ± 0.03	0.18 ± 0.02	0.10 ± 0.05	0.85 ± 0.1
Te	0.28 ± 0.03	0.18 ± 0.02	0.10 ± 0.05	0.85 ± 0.1
Si	0.43 ± 0.05	0.33 ± 0.05	0.10 ± 0.05	1.25 ± 0.1
Sn	0.19 ± 0.02	<0.1	0.10 ± 0.05	1.1 ± 0.1
GaAsP				
S	0.35	0.15 ± 0.03	0.20 ± 0.03	1.53
Te	0.19 ± 0.02	0.12 ± 0.03	0.07	0.65 ± 0.05

Source: Lang (1992).

ionization near the threshold was explained by the inferred densities of vibronic states that participate in the transitions. The relatively large values of E_0 for ordinary donors in these alloys (a later value of 160 meV was found for Si in AlGaAs) explain the problems that were found in effectively doping them in device applications using $x > 0.22$ in $Al_xGa_{1-x}As$, as indicated in Figure 2.1. The donor atoms are present, but their electrons are localized in the deep levels at ordinary temperatures, so they fail to produce the conductivity for which they were introduced. Because doping limitations occur frequently in II–VI compounds and DX-like behavior is seen in them also, this behavior provides a common thread between the two sets of materials.

These observations made a strong case for a LLR model for the DX center. There have been more recent attempts to invoke small lattice relaxations (Henning and Ansems 1987), but these seem to have been effectively refuted (Northrop and Mooney 1991).

2.3 Effects of Pressure and Alloying

Despite the many successes of the LLR model in providing a reliable framework for further research, its acceptance was slowed by the usual difficulties of any new model, and it could not identify the microscopic nature of the DX center. The familiar energy-level diagrams as a function of position are not amenable to the inclusion of lattice-relaxation effects, and the LLR approach created conceptual problems for some workers who exclusively used one-electron energy bands to describe the gap states of defects. That problem was the basis of attempts to explain energy barriers as being due to Coulombic ef-

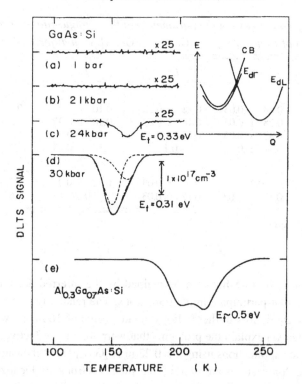

Figure 2.5 DLTS spectra for GaAs:Si for different pressures (a)–(d) as well as for (e) Al$_{0.3}$Ga$_{0.7}$As:Se. For (d) the observed DLTS curve is much broader than the calculated one, and dotted curves are calculated DLTS spectra for two well-defined traps required to describe the experimental curve. The inset shows a proposed configuration-coordinate model in terms of states associated with different conduction band minima near $x = 0.3$. *Source:* Mizuta et al. (1985).

fects (Craford et al. 1968). Also, there was a widespread conviction that complexities of the conduction band of AlGaAs were involved in the DX center. In the course of developing the picture further, detailed studies of the effects of hydrostatic pressure and alloying played crucial roles. The reasons for their importance are that they both affect energy-band structures in known ways, and it was gradually found that their effects on the properties of DX centers generally disagreed with band-structure effects.

The first clear evidence that the DX center is just a donor in disguise, rather than a complex formed by a donor associated with another defect such as a vacancy, came from measurements of the pressure dependence of DLTS properties of GaAs:Si (Mizuta et al. 1985). It was observed that a DLTS peak corresponding to the DX center appeared reversibly in unalloyed GaAs:Si when

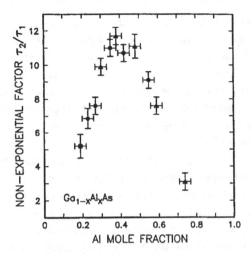

Figure 2.6 Variation of the nonexponential factor associated with the thermal emission of electrons from DX centers in $Al_xGa_{1-x}As$:Si as a function of the Al mole fraction x. *Source:* Calleja et al. (1986).

a critical pressure of ≈ 25 kbars was exceeded, as shown in Figure 2.5. The strength of the peak indicated that the density of the centers was comparable with the total donor density ($\approx 10^{18}$ cm^{-3}). This ruled out vacancies as participants because it is known that they do not come and go with pressure in such densities. It was also found that alloying with AlAs reduced the critical pressure, and at 30 percent Al the DX peak appeared without pressure, as seen in the lowest curve of Figure 2.5. The occurrence of the defect in GaAs also eliminated a role for Al in its formation, which had been conjectured. In attempting to explain these results, however, a purely electronic model based on pressure and alloy effects on the band structure was proposed (Mizuta et al. 1985); such models lack the LLR character and, as we show in Section 2.5, cannot explain the remarkable kinetics of DX centers. Nevertheless, this appears to be the first identification of the shallow–deep conversion of donors to form DX centers.

Another significant step was taken in an elegant set of experiments that utilized the nonexponential character of transients in thermal emission of electrons from DX centers in Si-doped AlGaAs. The nonexponential transient was approximated by expressing it as the sum of two simple exponentials, a fast and a slow component, with time constants τ_1 and τ_2. An index of the departure of a transient curve from a true exponential form was taken to be the ratio τ_2/τ_1, which is unity for a simple exponential. It was found that this non-exponential factor varies markedly with alloy compositions, as shown in Figure 2.6 (Calleja et al. 1986). This was interpreted as demonstrating that *alloy broadening* is responsible for observed properties of DLTS lines, since the presence of more than one time constant is known to broaden such lines. The binary-compound limits for both GaAs and AlAs have simple-exponential transients. Such studies of the kinetics of transitions between the two states of metastable centers have been powerful tools in their elucidation; they are dis-cussed further in Section 2.5, as well as in connection with defects in II–VI compounds, and especially in a-Si:H.

Because of the many proposals that assigned the properties of the DX cen-ter to energy-band complexities and the known effects of pressure on the electron energy-band structure, the pressure dependence of the nonexponen-tial factor for several alloy compositions was also reported (Calleja et al. 1986). The finding was that pressure has no effect on the nonexponential fac-tor at any composition, thus effectively excluding any role for conduction-band complexity in this behavior. This convincing set of experiments had a significant part in laying to rest electronic models for the DX center, and it showed that the local environment of the donor atom (as influenced by the proximity of alloy atoms) is the dominant influence in the nonexponential character of DX-center kinetics.

One feature of the paper by Calleja et al. (1986) that has not received much attention is their report that when electron emission is stimulated by light in-stead of thermally, the transients are simple exponentials for any composi-tion, that is, $\tau_2/\tau_1 \approx 1$. The interpretation they offered at the time was super-seded by later work of theirs (Mooney, Theis, and Calleja 1991) showing that, for the case of optical excitation, the expected deviation from an exponential transient is too small to be detected.

The importance of alloy-induced disorder increased with more detailed measurements of DLTS, finding that the alloy-broadened DX peak could be resolved into several components whose relative strength changes as the alloy composition changes (Mooney, Theis, and Wright 1988). This has been con-firmed subsequently in several ways (Baba et al. 1989; Brunthaler and Kohler 1990), one of which used persistent photoconductivity with results shown in Figure 2.7 (Brunthaler and Kohler 1990). As the Al content increases, first one, then two, then three peaks appear; complete studies have found altogeth-er four peaks, and no more. These are now accepted to represent the effects of different numbers of Al atoms in the neighborhood of a donor atom, and the composition alters the statistical probability of having any particular number of Al neighbors for a donor. It is significant that at most four peaks are ob-served, since these correspond to DX centers having 0–3 nearby Al atoms. These effects are small, however, because for AlGaAs:Si both the donor Si

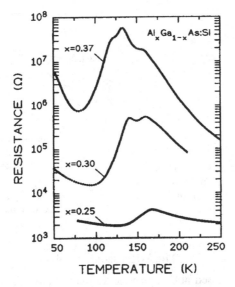

Figure 2.7 Resistance of $Al_xGa_{1-x}As$:Si samples with $x = 0.25$, 0.30, and 0.37, during heating after illumination. *Source:* Brunthaler and Kohler (1990).

and the Al alloy atom reside on Ga sites, so an Al atom cannot be closer than a second neighbor to a donor.

These four types of environments for a donor lead to four quantitatively distinct values of DX properties, and hence to four possible energy levels for an electron in one of the defects. Centers with more Al neighbors have larger electron binding energies. Their relationships to the energy-band structure of AlGaAs have been evaluated and are displayed in Figure 2.8 for Al concentrations up to 60 percent (Mooney 1991). The binding energy of an electron to the donor atom is 7 meV for $x < 22$ percent and then rises steadily to ≈ 160 meV as x increases to 40 percent. Figure 2.8 attempts to indicate the composition ranges in which each of the four types is clearly visible by showing them as solid lines, whereas dotted extensions show compositions in which a given type becomes undetectable. The previously noted broadening of DLTS lines is now recognized to represent sums of the contributions from these four possible variants of the DX center in an alloy. From detailed measurements on a set of dilute Al alloys of GaAs, Mooney and coworkers concluded that the proximity of an Al atom (i.e., a second neighbor) causes a group IV donor (Si) to move toward the interstitial site near the Al atom, thus changing the properties of the center (Mooney, Theis, and Calleja 1991). These results reinforced the conclusion that the properties of the DX center are dominated by configurational changes and not effects of conduction-band valleys.

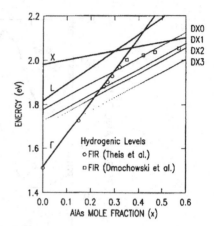

Figure 2.8 Band edges and important deep and shallow levels in $Al_xGa_{1-x}As$ as a function of alloy composition. *Source:* Mooney (1991).

2.4 The Microscopic Model

All the results described above provide so many constraints on acceptable atomic models for the DX center that its nature finally became generally evident in 1988–9. Purely electronic models such as Coulombic effects or band-structure effects had all been eliminated, and large lattice relaxation was required. In retrospect, the solution of this problem seems conceptually simple, but it needed the introduction of an unfamiliar condition: two relatively stable sites for a donor atom with nearly the same energies. Only by means of a thorough quantitative calculation could such a model be established.

The generally accepted atomic model is the *displaced-atom model,* in which a donor atom can occupy either of two sites: the usual substitutional, four-coordinated site or an adjacent interstitial site where it is three-coordinated. This model is consistent with an early proposal by Kobayashi, Uchida, and Nakashima (1985), one by Morgan (1989), and other suggestions (Mooney et al. 1991). It was Chadi and Chang (1989b) who performed the necessary calculation, an ab initio pseudopotential calculation of the energies of a wide range of atomic configurations, showing quantitatively that the two configurations in Figure 2.9 are the stable ones. This model explains essentially all the known properties of the DX center. The nature of the previously hypothesized lattice relaxation is finally clarified, and the shallow-or-deep-level character of the two associated electronic states is an integral feature. Although this behavior is that of an isolated donor alone, the picture of the defect state it presents is sometimes described as a V_{Ga}–D_i pair, that is, a vacancy at a Ga site paired with a donor at a *neighboring* interstitial site. Although the term *interstitial site* is often used for this purpose, the detailed calculations do not always place the donor at precisely the normal interstitial location.

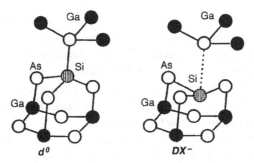

Figure 2.9 Stable configurations for a Si-donor impurity in GaAs. *Source:* Chadi and Chang (1989b).

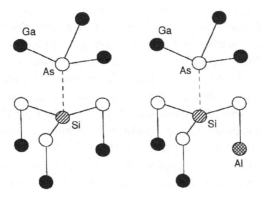

Figure 2.10 Effect of an Al neighbor on a Si-donor impurity in AlGaAs changes the energy of the deep-level configuration shown in Figure 2.9. There may be one, two, three, or no such neighboring Al atoms.

In the substitutional fourfold configuration the donor atom has the normal properties of a shallow donor with a weakly bound, large-orbit, effective-mass, Coulombic state for an electron. In the threefold configuration the donor has a localized electronic state whose binding energy is much stronger, thus forming the DX center. Despite the deep level, pressure can reverse the order of the energies of the two states because of the quantitative changes it causes in reducing the atom spacing, and alloying with Al produces a similar reduction. Proximity of an alloy atom to a donor, as shown in Figure 2.10, alters the quantitative details and the energy of the deep level relative to the unalloyed case. The possibility of up to four different environments for the defect in alloys accounts for the fine structure in DLTS and in persistent photoconductivity, and their sum provides the nonexponential kinetics in alloys that does not occur in binary compounds.

Whenever the ground electronic state of such a center happens to be the localized, deep-level state, thermal energy may be insufficient at room temperature to free the tightly bound carriers. This explains the difficulty in achieving high conductivities in some n-doped AlGaAs alloys, and Chadi and Chang (1989a) found similar results for p-type, As-doped ZnSe. As discussed in the next chapter (Section 3.2), this mechanism provides a major contribution in limiting the ability of many II–VI semiconductors to be doped.

For DX centers formed by column-VI donors in AlGaAs, which normally reside on As sites, the details of the reconfiguration are slightly different. In the case of S, the defect is formed not by motion of the donor atom, but instead by motion of one of its neighboring Ga atoms to an interstitial position (Chadi and Zhang 1991). There do not appear to be any other qualitative differences in these cases from those of column-IV donors on Ga sites.

One unexpected feature of the Chadi–Chang calculations of the energetics of DX centers in AlGaAs alloys is the conclusion that they have *negative U,* that is, negative effective electron correlation energies. This was described in Section 1.8 as a property of the vacancy in Si and of defects in chalcogenide glasses, and now seems to prevail in the DX centers. Hence, a change of state of such centers requires participation of *two* electrons, not one. Chadi concludes that the defect state (the state with threefold coordination) occurs only when the donor has a negative charge; this is the explanation for the apparent absence of an ESR signal from DX centers (Chadi and Chang 1989b). There has been some difficulty in gaining acceptance of this aspect of the microscopic model, perhaps because a donor with a negative charge is unfamiliar; but, of course, in the threefold state the electronic wave functions are localized, and the donor acquires many other new properties as well, so a negative charge need not be rejected. Because formation of the DX state requires it to have two electrons, only half of the donors can make the transformation; this is represented by the relation

$$2d^0 \rightarrow d^+ + DX^- \tag{2.1}$$

which indicates that two neutral substitutional donors with their electrons lead to one normal (i.e., fourfold), positively charged donor and one negatively charged DX center. Thus they effectively compensate each other, and the electrons in the DX center are generally bound too tightly to provide conduction at normal temperatures.

Although this displaced-atom model is widely favored now, the debate is not over. An independent analysis of two competing theoretical calculations was recently undertaken by Baraff (1991), who concluded that the claims in favor of these others had not been supported adequately, although they each had some merits of their own. A similar attempt by Lannoo (1991) led to somewhat similar results. Also at that same conference, Morgan (1991) proposed still another model in which the DX center in GaAs:Si is formed by the exchange of sites of Si_{Ga} and a host As atom, thus forming an As_{Ga}–Si_{As} pair.

A brief digression on terminology is in order here; it will become more significant in discussing metastable defects in a-Si:H. A photoinduced change of state of a donor between its fourfold site and the threefold site seems phenomenologically to be *generation* or *destruction* of a DX center, since a deeplevel, localized defect appears or disappears where there had been only an ordinary donor with no localized properties. Thus, although it may appear as if a defect is created, in reality only a change of state occurs, and it is only defect *properties* that are generated.

2.5 Transition Kinetics

In experimental studies that helped to characterize the DX center and lay the foundation for its understanding, kinetics of the transitions between the two states have been very important; they are equally important in studies of II–VI compounds, and particularly in a-Si:H. Persistent photoconductivity, which was perhaps the initial evidence of unusual properties, is itself a kinetics observation. From the early observations of the nonexponential character of relaxation of PPC (Lang and Logan 1977; Nelson 1977) and the use of a "nonexponential factor" (Calleja et al. 1986) whose properties are shown in Figure 2.6, these transients have contributed to the evolving picture of the DX center in III–V compounds. More recently these transients have been well fitted (Campbell and Streetman 1989) with *stretched exponentials* (SEs), which have the form

$$N(t) = N_{sat} - (N_{sat} - N_0) \exp[-(t/\tau_0)^\beta] \qquad (2.2)$$

where $N(t)$ is the density of defects as a function of time, β is the *stretch parameter* ($\beta < 1$), τ_0 is an effective time constant, N_0 is the initial value, and N_{sat} is the steady-state value (after times long compared to τ_0). It is clear that as β approaches 1 this form approaches a simple exponential. It is well known that SEs occur as the result of distributions of values of the physical parameters in kinetics, and that many SEs can be approximated by a sum of simple exponentials. These topics are discussed in more detail in connection with the kinetics of defects in a-Si:H in Section 5.1.

The SE fits, and the alloying behavior shown in Figure 2.6, might be interpreted as sums of simple exponentials, each of which represents one of the four possible numbers of foreign-atom neighbors as discussed in connection with Figure 2.10. However, in detailed kinetics studies of both relaxation and buildup transients of PPC in $Al_{0.3}Ga_{0.7}As$ and $Zn_{0.3}Cd_{0.7}Se$ alloys, SEs were found to describe all cases (Dissanayake et al. 1992). In the AlGaAs material, which is well characterized, Dissanayake et al. (1992) find that the SE is a better fit than any four-term sum of simple exponentials (H. Jiang, private communication). One explanation of this may be that effects of compositional fluctuations, which were identified as important in DLTS properties of InGaP and InGaAsP (Yoshino et al. 1984), in $GaAs_{1-x}Sb_x$ (Diwan et al. 1987), and dominant in the kinetics of ZnCdSe (Dissanayake et al. 1992), might play an additional role. In such situations there are two types of contributions to the character of transients: one from the simple sum of distinct microscopic variants of the defect, and the other from potential fluctuations that inevitably occur in inhomogeneous materials. Because both of these vanish in simpler binary compounds, it is not yet clear how to separate them.

2.6 EL2 and Other Defects in III–V Compounds

The defect designated EL2 is important in GaAs because it is an intrinsic defect that appears in undoped materials and has significant technological impact as the main compensator in semi-insulating material. It is metastable, with properties that resemble those of the DX center (although EL2 is more complicated) and it is now believed to have an origin related to the displaced-atom model of the DX center (Chadi and Chang 1988; Dabrowski and Scheffler 1988, 1989). Many of the observed properties of EL2 have been summarized together with an extensive set of references (Baraff 1992), so this discussion is limited to a brief summary of its major features.

The central fact of the EL2 model is that it is an As_{Ga} antisite defect. This misplaced As, however, can occupy either of two sites (just as with a donor that forms a DX center) that are of nearly the same energy, thus forming a metastable center. One is the normal Ga substitutional site with fourfold, tetrahedral coordination; the other is an adjacent interstitial site with threefold coordination, which may be stated as $As_{Ga} \leftrightarrow V_{Ga}As_i$. (It is an interesting historical coincidence that the two independent calculations that reached this same conclusion for the microscopic description of the EL2 appeared as adjacent articles in the same publication [Chadi and Chang 1988; Dabrowski and Scheffler 1988].)

The terminology that has come into common use refers to the simple antisite defect (i.e., an As atom at a substitutional Ga site) as just EL2. In this normal, fourfold state it is believed that the EL2 center may be neutral ($EL2^0$) or singly or doubly positively charged. Thus EL2 can have two midgap levels

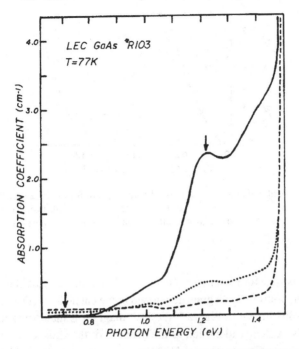

Figure 2.11 EL2 absorption spectra at 77 K for GaAs. The solid line is obtained after cooling in the dark before quenching. The dashed line is obtained after long exposure to a white light source. The dotted line is partial recovery obtained by 60-min irradiation with 0.90-eV photons following total quench. Arrows indicate points where absorption coefficient was monitored with time. *Source:* Fischer (1987).

corresponding to 0/+ or +/++ charge-state transitions. These charge states permit EL2 to be an effective recombination center for carriers, with impacts on device properties. The metastable (EL2*) state, corresponding to a threefold coordination, is very difficult to detect at normal pressures and seems to be resonant with the conduction band. One of the key properties of $EL2^0$ is that in its normal state it has an optical absorption band at 1.2 eV that can be bleached (also called *quenched*) by exposure to light, which excites the centers into their metastable state ($EL2^0 \rightarrow EL2*$); this behavior at 77 K, as reported by Fischer (1987), is shown in Figure 2.11. Relaxation from the EL2* state to $EL2^0$ occurs at temperatures above ≈ 120 K with an activation energy of 0.36 eV. The EL2* state does not absorb light, so the transition EL2* $\rightarrow$ $EL2^0$ cannot be induced optically, an observation that has received both theoretical and experimental attention.

Among the conclusions for which there seems to be a consensus is that higher excited electronic states in both configurations must play some roles

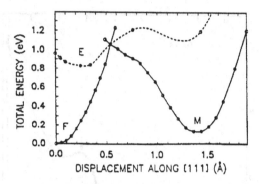

Figure 2.12   Calculated configuration-coordinate diagram for the EL2 defect. The co-ordinate Q is the distance that the As atom is from the substitutional Ga site. The upper curve is for an electronically excited state. *Source:* Dabrowski and Scheffler (1989).

in the various transitions. These states are included in the rather complicated configuration-coordinate diagram developed in the calculation (Dabrowski and Scheffler 1989) shown in Figure 2.12. That diagram is interesting in that the coordinate is explicitly identified as the distance of the antisite As atom from the substitutional Ga site along a [111] direction. In the metastable configu-ration it is 1.4 Å, a very large displacement that means that one of the tetra-hedral bonds to the atom is effectively broken. Despite its complexity, this type of CC diagram is significant because it represents one of several bond-breaking defect processes: EL2, DX, and the dangling-bond defect in a-Si:H.

Hydrogenation of GaAs is known to produce passivation of dopants by forming small complexes with them, and after acceptors in p-type GaAs have been passivated by hydrogen it is possible to reactivate them optically (Sza-franek, Bose, and Stillman 1989). The activation process is assumed to be the dissociation of the hydrogen-passivated complex followed by diffusion of the hydrogen away from the acceptor. This reactivation with above-band-gap light can occur at temperatures as low as 1.6 K and seems therefore to be athermal. Other workers have found metastable behavior of new deep levels in hydro-genated GaAs:Si at 0.60 eV below conduction-band energy E_C (Cho et al. 1991).

Surprising properties have been reported in GaAs:Ge, which appears to be either semimetallic or semiconducting depending on pressure and previous il-lumination (Skierbiszewski et al. 1993). Among other interesting properties are the "drastic enhancement" of the electron mobility under conditions of persistent photoconductivity, although the usual behavior (see Section 2.1) is a decrease in mobility.

Metastability has been observed in several types of defects in InP. In electron-irradiated ($\approx$1 MeV) n-type InP, a defect called the *M center* arises that has been interpreted as having two possible configurations (Stavola et al. 1984). Although reported before the current models of DX centers were developed, M centers were thought to have many of the same properties, including negative *U*. Somewhat different but metastable properties reported in un-irradiated Fe-doped InP have been designated the MFe center (Levinson et al. 1984). A DX center has been reported in InP:S under pressure greater than 82 kbar; its properties were described with the same formulation now used for the DX center in GaAs under pressure (Wolk 1992).

A number of other metastable effects have been observed in other semi-conducting compounds. Persistent photoconductivity has been reported in nitrogen-doped 6H-SiC at low temperatures (Dissanayake and Jiang 1993). The results were interpreted as being due to a center that undergoes lattice re-laxation like the DX center in AlGaAs, although the data indicate a barrier for electron capture of only 15 meV (cf. 160 meV in AlGaAs). Thermal relaxation thus occurs at much lower temperatures: At 10 K the stretched-exponential fitting curve has a time constant of 3,400 s. Because of the complicated struc-ture of 6H-SiC, however, little could be inferred about the microscopic origin of this metastable effect.

2.7 Persistent Optical Quenching of Photoconductivity in GaAs

Since persistent photoconductivity has played such a characteristic role in de-scribing the properties of photoinduced defects, it is worthwhile to note also the possibly related phenomenon of persistent optical quenching of photocon-ductivity in high-resistivity GaAs crystals.

In the relatively early days of research on GaAs, it was found that the in-clusion of oxygen gave rise to high-resistivity material, indicating the exis-tence of deep levels (Gooch, Hilsum, and Holeman 1961; Haisty, Mehal, and Stratton 1962; Woods and Ainslie 1963). If the spectral response of photocon-ductivity in such a typical crystal with room-temperature resistivity of 1.6×10^7 Ω-cm is measured above about 105 K, normal behavior is found with a maximum near the band gap at 1.4 eV and extrinsic response extending out to about 0.6 eV. When spectral response of photoconductivity is measured at 82 K, however, strikingly different results are found, as shown in Figure 2.13(a,b) (Lin, Omelianovski, and Bube 1976). Time constants in the critical photon-energy range (1.0–1.3 eV) exceeded several hours.

The results can be interpreted as indicating two distinct "states" of the crystal: a higher-sensitivity n-type state if the sample has not been previously

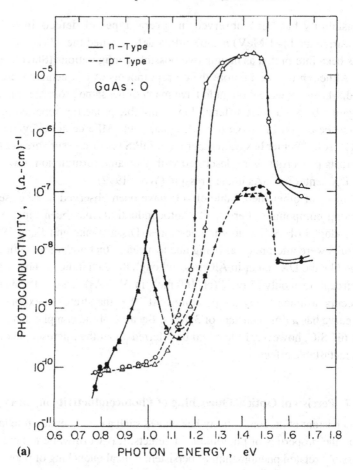

(a)   PHOTON ENERGY, eV

Figure 2.13 (a, *above*) Photoconductivity spectral response curves measured at 82 K for a high-resistivity GaAs crystal, previously heated and cooled (in the dark), measured from high to low photon energies, after 3 min (O) and after 15 min (△); measured from low to high photon energies, after 3 min (●) and after 15 min (▲). (b, *facing*) Photo-Hall mobility vs. photon energy for the same crystal at 82 K. Symbols are the same as in (a). *Source:* Lin, Omelianovski, and Bube (1976).

exposed to 1.0-1.3 eV photons, and a lower-sensitivity *p*-type state (about two orders of magnitude smaller than in the *n*-type state) to which the higher-sensitivity *n*-type state is transformed by exposure to photons in the 1.0–1.3-eV range.

The temperature dependence of photoconductivity for intrinsic photoexcitation (between 1.4 and 1.5 eV) and the corresponding photo-Hall mobilities are shown in Figure 2.14(a,b). Curve 1 represents the crystal in the high-

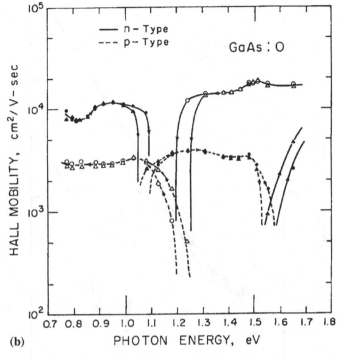

(b)

Figure 2.13 *(cont.)*.

sensitivity state, and curves 2 and 3 represent the crystal in the low-sensitivity state at 82 K. The optical quenching of photoconductivity is persistent at low temperatures, and the low-sensitivity quenched state remains unchanged long after the removal of the quenching radiation even with intrinsic photoexcitation present. Upon heating, however, a rapid and dramatic annealing from the low-sensitivity quenched state to the high-sensitivity state occurs when the temperature exceeds 105 K.

A variety of other investigations have also described the persistent photoquenching effect in *n*-type GaAs, with a general implication that it is the EL2 defect that is involved. These investigations have involved measurements of photocapacitance (Mitonneau and Mircea 1979), infrared absorption (Martin 1981), photoluminescence (Yu 1982), and the optical response of electron spin resonance (Tsukada, Kikuta, and Ishida 1985). In a recent investigation involving photocapacitance measurements (Nishizawa, Oyama, and Dezaki 1994), the authors report three ionized levels at 0.50, 0.66, and 0.74 eV above the valence band only during the photoquenching phenomenon. Each level changes in density during the photoquenching.

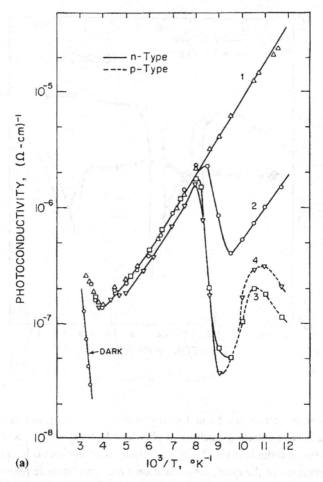

(a)

Figure 2.14 (a, *above*) Temperature dependence of photoconductivity for the high-resistivity GaAs crystal of Figure 2.13 for intrinsic photoexcitation after quenching 1 h with 1.08-eV (○) or 1.18-ev (▢) photons at 82 K, measured while warming; and for simultaneous intrinsic and 1.18-eV photoexcitation (▽). (b, *facing*) Temperature dependence of the photo-Hall mobility for the same crystal. Symbols are the same as in (a). *Source:* Lin, Omelianovski, and Bube (1976).

These phenomena, like many of the other photoinduced defect effects and their corollaries, offer a challenge to unravel the complexity of the observed behavior. In the present case one could propose two types of possible models for the low-temperature persistent optical quenching of photoconductivity that is rapidly thermally annealed above 105 K:

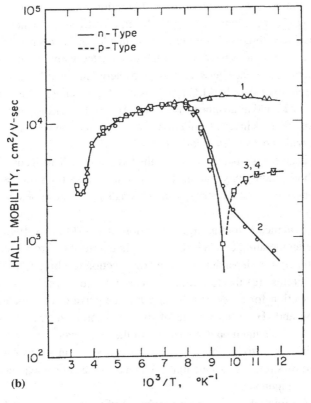

Figure 2.14 *(cont.).*

1. rearrangement of the occupation of existing defects with suitable capture cross sections so that large cross-section centers become empty at low temperatures and remain empty until the corresponding carriers trapped at other defects are thermally released to return; or
2. the photoinduced formation of defects with large capture cross sections that are thermally annealed above 105 K.

2.8 Summary of Properties of the DX Center in III–V Compounds

The properties of the DX center in III–V compounds (with a number of similarities in other compounds as well) can be summarized as follows:

1. The DX center is associated with isolated donor atoms rather than defect complexes, but it is not a state of a substitutional donor.

2. These donors have two possible configurations: The normal tetrahedral bonding gives the familiar shallow-level, effective-mass state; the other is a displaced threefold configuration with a localized, deep-level state.
3. These configurations have a metastable character, with a small equilibrium energy difference between their two energy minima.
4. The deep-level state in $Al_xGa_{1-x}As$ is at a higher energy than the other for $x < 0.22$ and atmospheric pressure. At $x > 0.22$ or high pressure the relative energies invert. Limitations on achievable conductivity are associated with the deep-level condition.
5. There is an energy barrier between the two minima. This leads to persistent photoconductivity and unusual capture cross sections for electrons.
6. The thermal ionization energy is much smaller than the optical ionization energy.
7. Much evidence favors a large relaxation model (Mooney 1990, 1991): (a) the temperature dependence of the photoionization cross section; (b) the temperature-dependent capture cross section resulting in PPC at low temperatures; (c) the low radiative capture rate; (d) the large capture cross section for holes; (e) the apparent negative effective correlation energy; and (f) the narrowing of the x-ray lattice diffraction peak of $Al_xGa_{1-x}As$:Te upon photoionization of the DX center.
8. The kinetics of transitions between the two states in alloys can be described by stretched exponentials, which become simple exponentials in binary compounds.
9. There are four DLTS peaks for $Al_xGa_{1-x}As$:Si, but only one for GaAs:Si.
10. There appears to be a general consensus that the DX levels are characterized by a negative charge state (Mooney 1991). Evidence supporting this conclusion of the Chadi and Chang (1989a,b) theory occurs in: (a) the absence of EPR activity of the deep-level state; (b) DLTS measurements with hydrostatic pressure on GaAs:Si:Ge indicating that each Ge donor must trap more than one electron (Fujisawa, Yoshino, and Kukimoto 1990); (c) an analysis of Mossbauer spectra on GaAs:Sn under hydrostatic pressure indicating the trapping of either two or three electrons at a DX center (Gibart et al. 1990); (d) photoconductance measurements indicating that the photoionization of DX centers corresponds to the removal of two electrons (Dobaczewski and Kaczor 1991); and (e) analysis of infrared absorption in GaAs:Si under pressure, indicating that each DX level traps two electrons (Wolk et al. 1991).

Properties 3–6 are the principal bases of the case for a large-lattice-relaxation description of the DX center, as portrayed in Figure 2.4. Quantitative values of the parameters in this configuration-coordinate diagram for AlGaAs doped with Se, Te, Si, or Sn, as well as for GaAsP doped with S or Te, have been tabulated (Lang 1992) and are given above in Table 2.1. For the common case of $Al_xGa_{1-x}As$:Si, $E = 0.43$ eV, $E_B = 0.33$ eV, $E_0 = 0.10$ eV, and $E_n = 1.25$ eV. Properties 9 and 10 effectively exclude all purely electronic descriptions of the DX center, such as band-structure effects.

3

Other Crystalline Materials

3.1 CdS and CdSe

Descriptions of exotic effects as the result of photoexcitation of II–VI materials, particularly CdS and CdSe, can be found extending back to when Boer, Borchardt, and Borchardt (1954) reported temperature-dependent, slow decreases of photoconductivity with time of photoexcitation of CdS crystals. In this section we review some examples of these effects and the kinds of models that have been suggested. It must be admitted that genuinely authenticated models do not exist for many of these early photoinduced defect effects, and so we primarily call attention to the suggestions that investigators have made.

Apparently related effects were found by measurements of thermally stimulated conductivity depending on the temperature of photoexcitation before making the measurement, and in the effect of photoexcitation on impure CdS, particularly on CdS:Cu crystals showing both a decrease in bulk photoconductivity of the material and a reversible degradation of the junction collection properties of a Cu_xS/CdS heterojunction. The phenomenon of persistent photoconductivity has also been observed. In many cases decreases of luminescence with time under photoexcitation have been observed, paralleling the decrease in photoconductivity. A photoinduced defect interaction model for some of these kinds of effects has been described by Tscholl (1968).

Decrease in Photoconductivity with Time in CdS and CdSe

The experimental results at 100 °C of the original experiment of Boer and coworkers (Boer 1954, 1990 [p. 1101]; Boer, Borchardt, and Oberlander 1959) are shown in Figure 3.1. The suggested interpretation was that when the photoexcitation is started, the photocurrent rises initially with a rise time con-

50

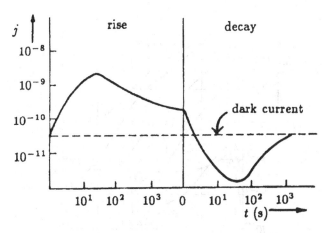

Figure 3.1 Rise and decay of the photocurrent for CdS at 100 °C. *Source:* Boer (1990).

trolled by the filling of traps, but then shows the effect of the photoinduced formation, with a much longer time constant, of recombination centers, which causes the photocurrent to go through a maximum (*overshoot*) and then decrease to a steady-state value at long times. When the light is turned off, the photocurrent decreases with a time constant controlled by the decay of free electrons, but then falls below the initial dark conductivity (*undershoot*). This result was interpreted to indicate that some donors have been eliminated in the step of photoinduced formation of recombination centers. With sufficient additional time, these photoinduced recombination centers anneal, the original donor density recovers again, and the dark current returns to its original value. Phenomenologically similar phenomena have been reported in CdSe for temperatures above –30 °C (Bube 1957).

These data illustrate one of the problems that sometimes attend the observation of apparent photoinduced defect effects; that is, Is the photoinduced defect interpretation indeed the correct one? In this case an alternative description of such overshoot and undershoot phenomena can be given in terms of a model involving rise and decay transients affected by traps and recombination centers without reference to photoinduced defect effects (Bube 1960/1978, p. 290). The observation of undershoot, for example, at the cessation of photoexcitation can be described in terms of a situation where the free and very shallow trapped electrons are drained off very rapidly by recombination, thus causing the rate of recombination to fall below the rate of electron escape from traps. When this occurs, the density of free electrons increases to the equilibrium value. Reasonable quantitative requirements can be formulated for capture

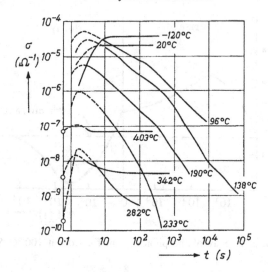

Figure 3.2 Variation of photoconductivity in CdS as a function of time for different temperatures at constant illumination (yellow Hg line). *Source:* Boer, Borchardt, and Borchardt (1954).

cross sections, Fermi-level location, and trap depth necessary to allow such an undershoot. Only a detailed investigation of all of the parameters can allow a choice to be made between these competing models.

Other data (Boer et al. 1954) on the long-term dependence of photoconductivity on time and temperature for measurements made at constant temperature and illumination are more indicative of general results that suggest photoinduced formation of defects. These are shown in Figure 3.2. Hall-effect measurements confirmed that the decrease in photoconductivity shown in Figure 3.2 is indeed due to a decrease in the free-electron density. The photoinduced-defect interpretation of the phenomena involves an initial photoexcitation that produces photoconductivity by exciting an electron from a defect, which then recombines again. However, the effect of photoexcitation is also to form new recombination centers, either by transformation of existing centers, or by the formation of new centers. For this to be effective the temperature must be high enough for thermal release of holes from the centers at which the initial photoexcitation occurred, followed by their capture at the new recombination centers.

Several related investigations describe the observed phenomena in CdS (and CdSe) in terms of the existence of two reversible states of the material: a high-photoconductivity $(\Delta\sigma_{\mathrm{hi}})$ state and a low-photoconductivity $(\Delta\sigma_{\mathrm{lo}})$ state (Bube 1959). Representative curves for the $\Delta\sigma_{\mathrm{hi}}$ and $\Delta\sigma_{\mathrm{lo}}$ states of CdS as a

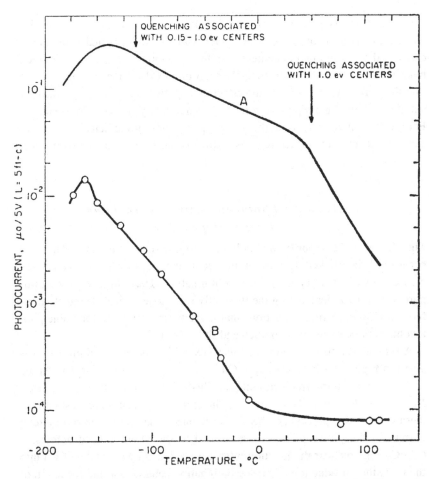

Figure 3.3 Temperature dependence of photoconductivity for special CdS crystals with a high-sensitivity state (curve A) and a low-sensitivity state (curve B). The CdS crystals were grown from the vapor phase and had iodine incorporated to make high-conductivity CdS:I crystals; these were then annealed for four hours at 900 °C under 2 atm of sulfur pressure to produce the high-resistivity crystals with reversible variations of photosensitivity. *Source:* Bube (1959).

function of temperature are given in Figure 3.3. The $\Delta\sigma_{hi}$ state is obtained by heating the crystals to 150–250 °C and then quenching. The $\Delta\sigma_{lo}$ state is obtained from the $\Delta\sigma_{hi}$ state by spontaneous decay at room temperature in either dark or light, greatly accelerated by light, or by heating the crystal to over 100 °C but below 150 °C; the $\Delta\sigma_{hi}$ state cannot be converted to the $\Delta\sigma_{lo}$ state by illumination at liquid-nitrogen temperature, however, indicating that a thermal contribution is required. It was suggested that a photoinduced-defect-agglomeration-dispersal phenomenon involving crystal defects or impurities may be involved. If the sensitizing centers in CdS are identified with Cd vacancies, the recombination centers with low ionization energies for captured holes may be associated with an agglomeration of Cd vacancies. The effect of illumination may be to reduce the effective negative charge on a Cd vacancy and thus to remove a repulsive force between similarly charged Cd vacancies. For the special crystals used, there is a higher-than-normal density of Cd vacancies expected in surface regions of the crystal because of the sulfur heat treatment.

Decrease in Photoconductivity with Time in Copper-Doped CdS

The effects of Cu impurity in CdS have been investigated in particular detail because of the marked decrease in photoconductivity in CdS:Cu crystals (as well as in CdSe:Cu crystals) due to illumination. These in turn play an important role in understanding the thermally restorable optical-degradation effects in Cu_2S/CdS heterojunctions that made them unsuitable for stable photovoltaic applications (Fahrenbruch and Bube 1974).

A typical variation of photoconductivity with time under illumination is given in Figure 3.4 for a crystal of CdS:Cu together with similar data on the variation of the short-circuit current for a Cu_2S/CdS junction (Fahrenbruch and Bube 1974). Research shows that both the optical-degradation process and the thermal-annealing process are thermally activated. The decrease in short-circuit current in the junctions can be attributed to the photoinduced defects in CdS:Cu regions near the junction interface produced by diffusion of Cu into the CdS, the introduction of fast recombination centers, and the reduction of the junction field in this region causing a loss in the collected short-circuit current.

From an investigation of the effects in CdS crystals with excess Cd donors and a density of Cu acceptors larger than that of Cd donors, Kanev and co-workers (Kanev, Sekerdzijski, and Stojanov 1963; Kanev, Stojanov, and Sekerdzijski 1964; Kanev, Stojanov, and Lakova 1969; Kanev, Fahrenbruch, and Bube 1971) have proposed a model for the photoinduced defect reaction involved. In CdS:Cd, it is assumed that Cu impurity forms two types of centers: Type-I centers are Cu-acceptor/Cd-donor pair complexes; type-II centers result from Cu impurities present in excess of the Cd concentration. Before optical degradation, the type-I levels are occupied and form neutral complexes with the Cd-donor levels, which are assumed to have a low electron-capture coefficient and a large hole-capture coefficient, thus giving rise to a high electron lifetime. Photoexcitation removes electrons from the type-I centers, plac-

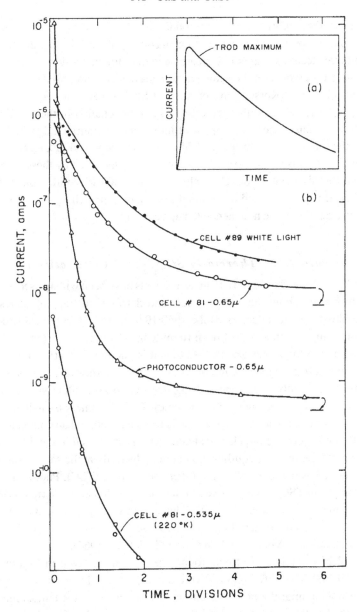

Figure 3.4 (a) Schematic thermally restorable optical degradation (TROD) curve showing the variation of short-circuit current of a Cu_xS/CdS junction with time after the beginning of photoexcitation. (b) Short-circuit current for a Cu_xS/CdS junction and photocurrent vs. time for a CdS:Cu photoconductor during optical degradation. Time scales for the four curves from top to bottom are 200, 4,000, 4,000, and 100 s/div, respectively. The upper three curves are for 300 K. Fully degraded values of current are indicated on the right-hand side for two of the curves. *Source:* Fahrenbruch and Bube (1974).

ing them on the type-II centers and producing type-I′ centers from the type-I centers. Since type-I′ centers can capture an additional hole, the capture coefficient for free electrons is greatly increased and their lifetime is decreased.

An alternative model for the same phenomena has been proposed by Sheinkman and colleagues (Korsunskaya et al. 1980; Sheinkman et al. 1982; Sheinkman 1987). They suggest that the effect of photoexcitation in these CdS:Cu crystals is to cause electron trapping at individual interstitial Cd_i, which results in the formation of clusters of $(Cd_i)_n$ that are deep-level fast recombination centers. This is the inverse of another process suggested by these authors in which the dissociation of $(Cd_i)_n$ clusters into shallow mobile donors Cd_i in CdS is strongly enhanced by infrared illumination that changes the cluster charge via capture of the hole generated at a cluster.

Variations in Thermally Stimulated Conductivity

Thermally stimulated conductivity measurements have been widely used to obtain information about the energy levels and densities of trapping states in high-resistivity semiconductors (Bube 1960/1978, p. 167). In a typical experiment, the sample is first cooled from room temperature to a low temperature either with or without photoexcitation; then it is photoexcited at the low temperature to fill trapping states, the photoexcitation is turned off, and the sample is heated, usually at a linear heating rate, while its conductivity is measured as a function of temperature (and time). The experiment is analogous to the "glow curve" in which luminescence is measured rather than conductivity. The difference between the electrical conductivity measured in the above experiment and the thermal-equilibrium dark conductivity at the same temperature is defined as the *thermally stimulated conductivity* (TSC). Photoinduced, slow changes in TSC curves were early detected by Woods and Wright (1960, p. 880), and observed variations in TSC have been widely attributed to photoinduced defects in the material in a continuing series of investigations (Woods 1958; Nicholas and Woods 1964; Woods and Nicholas 1964).

Typical of the kinds of effects found are the curves given in Figure 3.5 (Tscholl 1968). These TSC curves were obtained with illumination by white light during cooling; the photoexcitation was started when the temperature T was reached and then was turned off when a temperature of 0 °C was obtained. At liquid-nitrogen temperature the light was again turned on to fill the traps. Major differences among the curves are observed depending on whether or not photoexcitation occurred at higher temperatures. Other TSC curves – after illumination for different lengths of time at 50 °C before cooling without illumination to liquid-nitrogen temperature, where the crystal was again illuminated

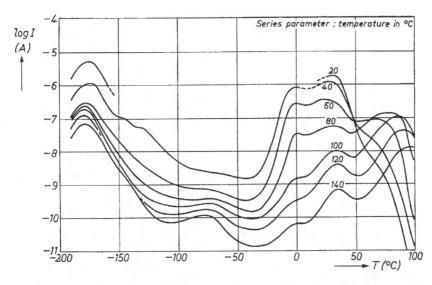

Figure 3.5 Thermally stimulated current curves for CdS after illumination with white light (no infrared) during cooling between temperature T °C (labeled "series parameter") given on the curves and 0 °C. *Source:* Tscholl (1968).

before subsequent heating in the dark – show similar variations in structure. It is evident that many complex phenomena are occurring in these crystals.

A particularly striking set of TSC effects in CdS and $CdS_{0.75}Se_{0.25}$ crystals has been reported (Bube et al. 1966; Im, Matthews, and Bube 1970), as well as similar effects in solution-sprayed CdS films (Chandra Sekhar et al. 1988). Figure 3.6 shows the TSC curves for these two crystals under two different sets of excitation conditions: (1) photoexcitation during the cooling cycle from high to low temperatures, and (2) photoexcitation at liquid-nitrogen temperature only. The TSC peak near room temperature is not found unless photoexcitation occurs during cooling at temperatures above about 180 K. An analysis of the TSC data shows that the particular peak in question near room temperature corresponds to a discrete trap with a depth of 0.73 eV that empties without retrapping although its emptying kinetics correspond to a rather large capture cross section of 1.3×10^{-14} cm^2.

A proposed mechanism for this effect involves the dissociation of a neutral center into a pair of shallow metastable electron and hole traps, both occupied; this dissociation has a maximum efficiency at a particular photon energy. The dissociated center forms a stable trap pair, either by lattice relaxation or diffusion (temperature-dependent processes), producing much deeper traps. When an electron is thermally freed from the electron trap, a hole is simultaneously released from the hole trap, resulting in the collapse of the center back

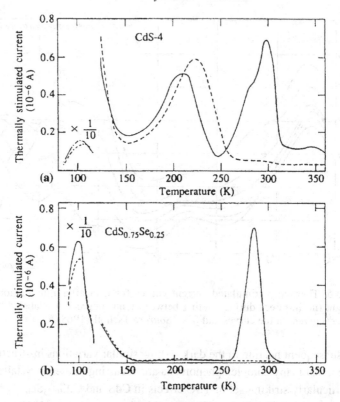

Figure 3.6 Thermally stimulated conductivity curves for (a) a CdS crystal irradiated with 240-keV electrons, and then annealed at 200 °C under photoexcitation,
and (b) a $CdS_{0.75}Se_{0.25}$ crystal. The solid curves indicate TSC measurements after
photoexcitation while cooling from high temperatures; the dashed curves indicate
TSC measurements after photoexcitation at low temperature only. *Source:* Im,
Matthews, and Bube (1970).

to its original configuration. Once the electron is excited into the conduction
band, the deep trapping levels no longer exist, and retrapping cannot occur.

Persistent Photoconductivity in CdS

Phenomena similar to persistent photoconductivity (PPC), often historically
called *stored conductivity* in CdS, have been observed at least since 1960, well
before a conscious encounter with either DX centers or dangling bonds, and
have been reviewed by Wright, Downey and Canning (1968). As in Figure
1.1 for AlGaAs, the effect consists essentially of the ability of light with a
photon energy of 1.76 eV to excite a metastable increase in dark conductivity

in CdS crystals at 77 K. Like other effects in CdS, the phenomena appear to be complex.

Three temperature regions are defined:

1. a low-temperature region (below ≈250 K) where conductivity storage can be excited by photoexcitation with photon energies greater than 1.76 eV;
2. a high-temperature region (above ≈350 K) where any existing storage is thermally quenched; and
3. between these two temperature regions, a region where the magnitude of the conductivity storage grows spontaneously without radiation.

The boundary temperatures vary between different samples. Hall measurements showed that the carriers responsible for the increased conductivity are electrons with a mobility of 7 cm^2/V s.

3.2 Other II–VI Materials

There is evidence that photoinduced defect interactions also occur in other II–VI compounds, such as CdTe and ZnSe, as well as in II–VI alloys.

Cadmium Telluride

An extensive literature exists dealing with phenomena in n-type CdTe that are similar to the phenomena involving DX-type defects in III–V compounds. Persistent photoconductivity has been observed both in undoped (Lorenz et al. 1964) and in heavily n-type doped CdTe (Iseler et al. 1972; MacMillan 1972; Losee et al. 1973; Baj et al. 1976; Dmowski, Porowski, and Baj 1978), and persistent cyclotron resonance has been reported for undoped CdTe (Pastor and Triboulet 1987). There has not been general agreement on the nature of the phenomena, although there is evidence that there may be more than one relevant mechanism, particularly indicating different defects involved in the undoped and n-doped materials. Earlier major models took one of the following two forms:

1. donor levels associated with other (X or L) conduction-band minima than the Γ minimum, following a pattern similar to that used in the description of the phenomena in III–V compounds before the development of the large-relaxation DX-center model, or

2. macroscopic barriers due to inhomogeneities in the material.

It now seems that a third model is likely:

3. DX centers in n-doped CdTe, like those in n-doped III–V compounds and n-doped ZnSe.

Effects in Undoped CdTe

Early results by Lorenz et al. (1964) concerning PPC in undoped CdTe were associated with a localized level lying 0.06 eV below the Γ minimum. They reported an activation energy for electron transfer of 0.2 eV, and proposed that the level was the second charged state of a double-acceptor center. Further data on this defect, labeled the *DA* (double-acceptor) *defect* in CdTe, are given below in comparison with the effects observed in *n*-doped CdTe.

More recent results have been described by Pastor and Triboulet (1987), who made measurements of persistent electron cyclotron resonance (PECR) in undoped CdTe at 4.2 K, as shown in Figure 3.7. After a rapid rise, the electron density indicated by PECR saturates at a value that depends on the photon energy used. When the photoexcitation is turned off, the electron density drops by several percentage points to a value that also depends on the photon energy used. An analysis of the whole phenomenon leads to the conclusion that additional electrons, freed by band-to-band absorption, participate in cyclotron resonance no different than for the nonpersistent case, and that PECR results simply from their slow recombination. Pastor and Triboulet propose a model based on macroscopic inhomogeneities in the CdTe as the mechanism by which photoexcited electrons and holes are separated by a potential barrier in the bulk of the material, thus making recombination difficult. Such a model has also been cited in several papers to explain persistent photoconductivity (Sheinkman et al. 1971; Vul and Shik 1973; Sheinkman and Shik 1976; Vul et al. 1976), and also for multilayer structures (Queisser and Theodorou 1979; Katalksy and Hwang 1984), where barriers arise naturally at the interfaces.

Effects in n-Doped CdTe

Detailed measurements of a variety of kinds have been made on *n*-type CdTe crystals doped with Cl or Ga (Iseler et al. 1972; MacMillan 1972). Similarities between the effects encountered with the two different donors indicate the common nature of the phenomena; differences indicate the particularities of each specific donor.

Conductivity and Hall Effect Two easily distinguishable photoconductivity processes have been identified and measured. In one case when $T < 200$ K, there is a significant increase in photoexcited electron density that is independent of temperature; the time constants are so long when $T < 150$ K that the photoconductivity persists for days. For this process, which is the process in which we are primarily interested here, there is little effect of photoexcitation on the mobility. For $T < 90$ K, a second photoconductivity process becomes

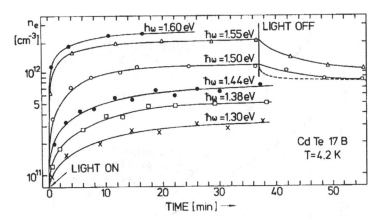

Figure 3.7   The rise and decay of persistent cyclotron resonance for different energies of exciting light in undoped CdTe at 4.2 K. The dashed line indicates a logarithmic decay. *Source:* Pastor and Triboulet (1987).

observable with a very slow decay ascribed by Lorenz et al. (1964) to DA-type defects, which is accompanied by a sizable increase in mobility. The energy level associated with DA-type defects lies about 0.06 eV below the bottom of the conduction band.

For the other defect, however, the experimental observation (MacMillan 1972) was that the measured value of $(n_L - n_D) = \Delta n$ is proportional to n_L^2, where n_L is the electron density in the light and n_D is the electron density in the dark. This led – through an analysis of the occupancy of the donors (invoking the relationships that $n_D = [D^+]$ and that $n_L = [D]$, where $[D^+]$ is the ionized donor density and $[D]$ is the total donor density) – to the conclusion that the donor level is defined at 150 K by $E_{Cl} = E_C + 0.029 \pm 0.008$ eV. Therefore the Cl-donor level appears to lie above the principal Γ minimum of the conduction band. Similar reasoning applied to CdTe:Ga crystals at 100 K led to the conclusion (MacMillan 1972) that at 100 K, $E_{Ga} = E_C + 0.014 \pm 0.001$ eV. Thus both Cl and Ga donors appear to have energy levels that lie above the energy of the Γ minimum, whereas the DA defect's level lies 0.06 below the energy of that minimum.

Decay of Photoconductivity   Measurements were made of the decay of photoconductivity as a function of temperature for CdTe:Cl, CdTe:Ga, and for DA defects in undoped CdTe corresponding to those above (Lorenz et al. 1964). Measurements of the exponential decay time constant associated with the DA defect as a function of temperature near 100 K give $\tau_0 = 3 \times 10^{-12}$ s and $E_B = 0.25$–0.27 eV (MacMillan 1972).

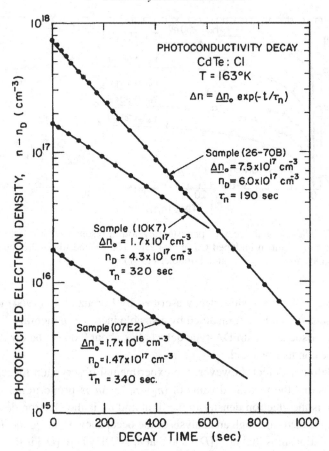

Figure 3.8 Decay of photoexcited electron density in CdTe:Cl crystals for various concentrations of Cl, resulting in differing densities for electrons in the dark and under photoexcitation. *Source:* MacMillan (1972).

For CdTe:Cl at 163 K, the decay of photoconductivity is exponential with a time constant of several hundred seconds, the time constant decreasing with increasing magnitude of initial dark electron density and subsequent photoexcited electron density (MacMillan 1972). Representative decay curves are shown in Figure 3.8. The temperature dependence of the exponential decay time constant was measured between 150 and 190 K for five different Cl concentrations. For all samples the time constant can be expressed as $\tau_n = \tau_0 \exp(E_B/kT)$, with $\tau_0 = 1.0 \pm (0.5) \times 10^{-13}$ s and $E_B = 0.51 \pm 0.01$ eV, a very large thermal activation energy. This result is surprising since the Cl-donor level appears to be close to the bottom of the Γ conduction band. When similar data for CdTe:Ga are analyzed, it is found that $\tau_0 = 1 \times 10^{-13}$ s and $E_B = 0.31 \pm 0.01$ eV.

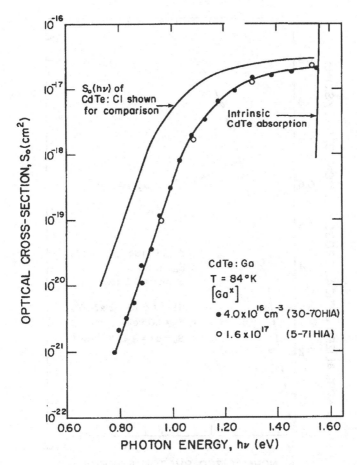

Figure 3.9 Optical cross section for a Cl-donor and a Ga-donor in CdTe, as determined from the spectral response of photoconductivity. *Source:* MacMillan (1972).

Optical Cross Section of Neutral Donors In order to compare optical ionization energies with thermal ionization energies, the spectral response of the optical cross section was measured at 84 K for both CdTe:Cl and CdTe:Ga, as shown in Figure 3.9 (MacMillan 1972). These optical cross-section spectra can be well expressed in terms of the model of Lucovsky (1965), for photo-excitation from deep levels assuming a delta-function potential for the ground state, with the results shown in Figure 3.10. The conclusion is that the optical threshold energies for photoionization of CdTe:Cl and CdTe:Ga are 0.90 and 1.03 eV, respectively.

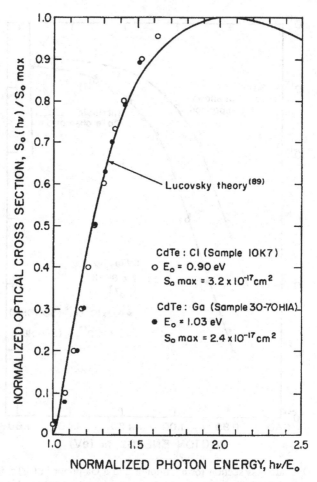

Figure 3.10 Comparison of optical cross-section spectra for CdTe:Cl and CdTe:Ga after replotting following the Lucovsky model with E_0 the optical threshold energy for the transition. *Source:* MacMillan (1972).

Hydrostatic Pressure Effects In experiments similar to those carried out in the research on DX centers in III–V compounds, Iseler et al. (1972) measured the effect of hydrostatic pressure at 300 K on the resistivity of CdTe with various different donors; the results are shown in Figure 3.11. This kind of effect has been described by Paul (Foyt, Halsted, and Paul 1966; Paul 1968) in connection with similar effects in III–V compounds (Feinleib et al. 1963; Zallen and Paul 1964). He attributed the increase in resistivity to an electronic band-structure effect.

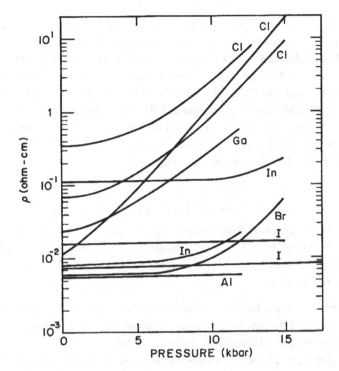

Figure 3.11 Dependence of the resistivity of *n*-type CdTe on the hydrostatic pressure at 300 K for different donor impurities. *Source:* Reprinted from *Solid State Communications* 10, 619, Iseler et al., "Non-*Γ* Donor Levels and Kinetics of Electron Transfer in *n*-type CdTe," © 1972, with kind permission from Elsevier Science Ltd., The Boulevard, Langford Lane, Kidlington OX5 1GB, UK.

Summary Both thermal and optical ionization energies associated with Cl and Ga donors in CdTe are much larger than would be expected for simple donors. A model was proposed in which each donor is predominantly associated with a different conduction-band minimum but neither is associated with the *Γ* minimum. In this model photoionization of a neutral donor is caused by optical excitation of the bound electron to the higher band minimum followed by phonon scattering to the *Γ* minimum. Recombination of free carriers with ionized donors occurs only for electrons with sufficient energy to occupy conduction-band states that overlap the ground state in *k*-space.

Because of the similarity between these results with those found and proposed earlier for persistent photoconductivity and related effects in III–V compounds, there appears to be a growing consensus that the DX-defect models now generally accepted for the III–V compounds are also appropriate for *n*-type

II–VI binaries such as CdTe (Legros, Marfaing, and Triboulet 1978; Takebe, Saraie, and Matsunami 1982; Debbag, Bastide, and Rouzeyre 1988; Khacha-turyan, Kaminska, and Weber 1989). A connecting investigation is described by Baj et al. (1976), who first showed the coupling between the electronic state of the defect and the crystal lattice in n-doped CdTe; pressure-induced changes in the electron concentration in CdTe:Cl were interpreted as indicat-ing that the donor ions can occupy two nonequivalent sites in the lattice sepa-rated by a potential barrier. Additional examples of n-type $Cd_{1-x}Mn_xTe$ and $Cd_{1-x}Mg_xTe$ are described in the following section.

Persistent Photoconductivity in II–VI Alloys

A number of researchers have reported the phenomenon of persistent photo-conductivity (PPC) in II–VI alloys, similar to that described for III–V alloys.

Three mechanisms have been proposed to account for PPC effects in a vari-ety of semiconductors.

1. In the microscopic local-potential fluctuation model, photoexcited carriers are separated by random local-potential fluctuations and thus prevented from recombining (Sheinkman and Shik 1976).
2. In doped GaAs-layered structures, the macroscopic barrier due to band off-set at the interface between layer and substrate is responsible for the carrier separation leading to PPC (Queisser 1985; Queisser and Theodorou 1986).
3. In the DX-center model, light excites electrons from deep states that un-dergo large lattice relaxation (LLR), resulting in small capture coefficients for the photoexcited carriers (Lin et al. 1976; Mitonneau and Mircea 1979).

In recent research on bulk materials, an investigation of PPC in ZnCdSe and CdSSe has been interpreted to support the first of these mechanisms (Jiang and Lin 1990; Dissanayake et al. 1991), whereas research on CdMnTe:Ga (Se-maltianos et al. 1993) and CdMgTe:Br or CdMgTe:Cl (Waag et al. 1994) has been taken to support the third. Here we summarize the principal features of these two investigations to illustrate the issues and results obtained to date.

Undoped II–VI Alloys

For an undoped $Zn_{0.3}Cd_{0.7}Se$ crystal, PPC could not be observed below 70 K, and the PPC relaxation is well described as a stretched exponential at $T <$ 220 K (Jiang and Lin 1990), with form

$$I_{PPC}(t') = I_{PPC}(0) \exp(-t'^\beta) \qquad (3.1)$$

where as usual $t' = t/\tau$, with τ being the characteristic stretched-exponential relaxation time constant. Typical PPC decay curves at 170 K are shown in

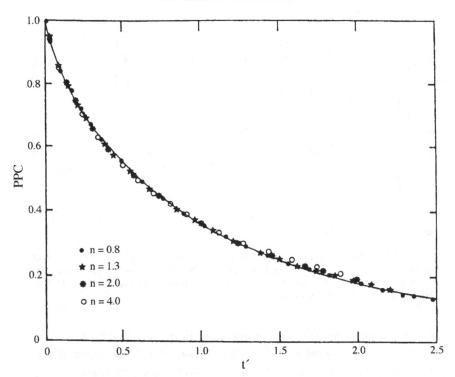

Figure 3.12 Normalized PPC decay curves obtained at 170 K for a $Zn_{0.3}Cd_{0.7}Se$ crystal after illumination with four different excitation photon doses. Each curve is normalized to unity at $t = 0$, the time at which photoexcitation ceased; the dark current has been subtracted out. The unit of the excitation photon dose n is 10^{16} photons/cm². The solid line is a plot of a stretched-exponential function with $\beta = 0.77$. The stretched exponential time constant τ increases with n, and is 604, 679, 734, and 776 s for the four values of n shown in the figure. The horizontal axis is $t' = t/\tau$, the scaled time. Source: Jiang and Lin (1990).

Figure 3.12 for different values of the photon excitation dose as a function of the scaled time t'. All the normalized decay curves can be well fitted by an SE with the decay exponent $\beta = 0.77$.

Furthermore, in the temperature range 125 K $< T <$ 220 K, the temperature-dependent growth of PPC after excitation with the same photon dose follows the percolation approach as

$$I_{PPC}(T, t=0) \propto (T - T_c)^{\theta} \tag{3.2}$$

with $T_c = 118$ K and $\theta = 1.3$. The temperature-dependent τ for the stretched-exponential decay also indicates a rapid variation near 120 K. Similar PPC behavior is found for a higher-quality $CdS_{0.5}Se_{0.5}$ alloy crystal, which exhibits

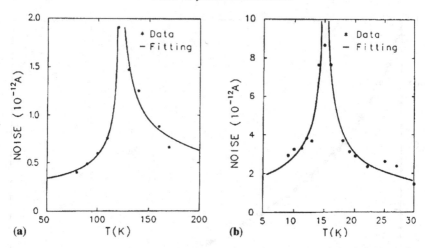

Figure 3.13 The magnitude of the measured PPC noise as a function of temperature for (a) a $Zn_{0.3}Cd_{0.7}Se$ alloy crystal, and (b) a $CdS_{0.5}Se_{0.5}$ alloy crystal. Solid lines are theoretical fits to the form of Eq. 4.3, with T_c given by 118 K in (a) and 15 K in (b). *Source:* Jiang and Lin (1990).

$T_c = 15$ K under the same experimental conditions, and even for a compensated CdS single crystal with $T_c = 1.5$ K, consistent with the expectation of much smaller potential fluctuations for the binary CdS crystal case.

In addition, measurements of conductivity noise in II–VI alloys follow a percolation description as $|T - T_c|^{-\theta}$ and reach a maximum at $T_c = 118$ K for $Zn_{0.3}Cd_{0.7}Se$ and at $T_c = 15$ K for $CdS_{0.5}Se_{0.5}$, as shown in Figure 3.13(a,b), respectively.

Additional measurements (Dissanayake et al. 1991) on the $CdS_{0.5}Se_{0.5}$ crystal of the variation of the PPC decay constants as a function of temperature, $\tau = \tau_0 \exp(E_{rec}/kT)$, give a value of $E_{rec} = 6.6$ meV for the recombination barrier height, indicating that a small recombination barrier can give rise to PPC. Measurements were also made of time-resolved photoluminescence of exciton transitions in this crystal. At 8.4 K the photoluminescence of the localized-exciton transition showed a full width at half maximum (FWHM) of 11.3 meV, some fifty times larger than is found in a good-quality binary CdS or CdSe crystal, thus indicating compositional disorder in the material.

These results have been interpreted in support of the following model invoking alloy-induced compositional fluctuations as the principal cause of the effect (Jiang and Lin 1990; Dissanayake et al. 1991). Photoexcited holes are localized at the low-potential sites in the conduction band; holes are localized at such sites in the valence band. Since the low-potential sites in conduction and valence bands occur at spatially separated sites in the crystal, recombination between photoexcited carriers is prevented and PPC results. Below T_c transport is by electrons hopping between local potential minima, producing a low-level PPC that becomes negligible at very low temperatures. For temperatures greater than T_c the electron sites form a percolation network and electron transport changes from localized hopping to percolation (delocalized) paths, so that the conductivity corresponds to electrons percolating through the network of available states. It is suggested that the scaling behavior seen in Figure 3.12 indicates a fractal structure for the network of percolation-accessible sites for electrons. The decay exponent β is independent of photoexcitation dose and therefore depends only on the fractal dimension. Neither the macroscopic barrier model nor the LLR-defect model appear to be consistent with the data, especially with the localized-to-percolation transport phase transition observed.

Doped II–VI Alloys

Different conclusions have been drawn from an investigation of PPC in crystals of $Cd_{0.97}Mn_{0.03}Te$ heavily doped with Ga donors (Semaltianos et al. 1993), suggesting that different mechanisms dominate in undoped and doped II–VI alloys. The results with the doped alloy are more consistent with those described in the previous section for doped CdTe binary crystals. Transport data obtained for $Cd_{0.97}Mn_{0.03}Te$:Ga as a function of temperature are given in Figure 3.14, clearly showing the existence of PPC for T < 60 K. (Note the similarity of Fig. 3.14b to Fig. 1.1.) If the electron density is expressed as $n = n_0 \exp(-E_d/kT)$, then the data indicate a thermal ionization energy of $E_d = 60$ meV for the donors. Below 105 K, the slope of the dark value of n vs. $1/T$ changes as the electronic equilibrium begins to lag behind the structural equilibrium, and at lower temperatures it becomes impossible to reach electronic equilibrium in a finite time. Photoexcitation at low temperatures excites electrons from deep levels to the conduction band; at these temperatures the electrons have insufficient energy to overcome the capture barrier and to return to the deep levels. Above 60 K, however, the electrons can surmount the barrier, and the density of free electrons decreases rapidly to the thermal-equilibrium value. This is essentially the same description as that given in Section 1.6.

Figure 3.15 shows the electron density growth kinetics at 40 K for different photon energies. The curves could not be fit by a simple exponential, and can be formally fit by a stretched exponential only if $\beta = 1.6$. It was claimed that the transients can be successfully modeled as the sum of two exponentials, which can be interpreted as arising from the electron density transient resulting from a two-step photoionization via an intermediate state of one type of defect.

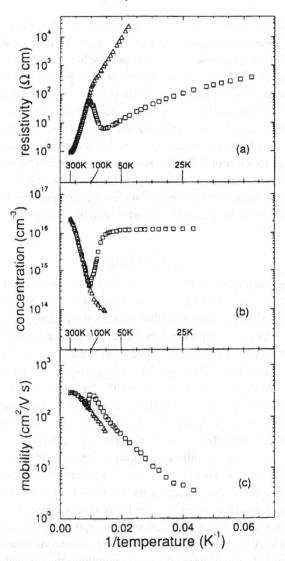

Figure 3.14 Temperature dependence of the resistivity, electron concentration, and electron mobility from Hall measurements on $Cd_{0.97}Mn_{0.03}Te:Ga$, showing persistent photoconductivity for $T < 60$ K. ($\triangle$) Data taken during slow warming in darkness; ($\square$) Data taken after exposing the sample to light at low temperatures. *Source:* Semaltianos et al. (1993).

A recent investigation of *n*-type doping of CdMgTe with bromine and chlorine prepared by molecular beam epitaxy has also been interpreted in terms of a DX-like state in the alloy, which moves into the band gap with increasing Mg concentration (Waag et al. 1994). In the absence of Mg, both Br and Cl

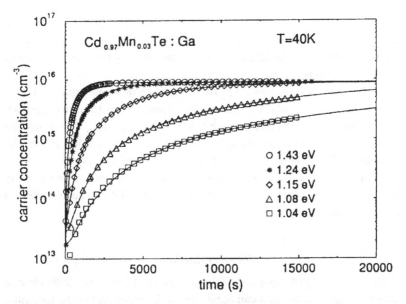

Figure 3.15 Electron density vs. time in $Cd_{0.97}Mn_{0.03}Te$:Ga at 40 K for different photon energies. Solid curves are obtained by a fitting process assuming a two-step photoionization model. *Source:* Semaltianos et al. (1993).

give rise to shallow donors corresponding to free-electron densities in the range of 10^{18} cm^{-3}. For increasing Mg concentration in the alloy, however, deep donors are created that limit the free-electron density at room temperature and can be converted by photoexcitation below 180 K into metastable shallow donors that give rise to PPC.

Finally, a recent investigation of transport in large single crystals of $Cd_{0.8}Zn_{0.2}Te$:Cl indicates that chlorine donors in CdZnTe form DX centers (Bennett et al., in press). Persistent photoconductivity has been observed with an annealing temperature of 130 K, and Hall-effect measurements indicate that the PPC arises from a persistent increase in the density of free carriers. At 100 K the conductivity in the saturated photoconducting state is 10^8 times that in the dark.

Summary

Dissanayake et al. (1992) have made a detailed comparison of PPC phenomena in doped $Al_{0.3}Ga_{0.7}As$:Si and high-resistivity $Zn_{0.3}Cd_{0.7}Se$ semiconductor alloys. They find that the PPC relaxation in both materials can be well described by stretched exponentials (SEs). They also measured the PPC-buildup transients, finding them to be quite different in the two materials. All their results are consistent with the conclusion that the behavior observed in the $Al_{0.3}Ga_{0.7}As$:Si is associated with the properties of DX centers, whereas the behavior observed in the $Zn_{0.3}Cd_{0.7}Se$ is governed by tail states related to compositional fluctuations.

That the behavior may also be a critical function of preparation conditions is also indicated by an investigation by these same authors (Dissanayake, Lin, and Jiang 1993) on PPC effects in *thin films* of $Zn_{0.04}Cd_{0.96}Te$ prepared by the laser ablation method. Once more the PPC in these materials can be described by an SE function with very long lifetimes; the results indicate, however, that the PPC in the films is associated with deep centers, presumably induced by the deposition technique, and possibly exhibiting both a stable and a metastable state.

Zinc Selenide

Because of its wide energy gap (2.8 eV), ZnSe is an attractive candidate as a photoelectronic material that can luminesce in the green part of the spectrum, and it has become the focus of much recent work aimed at producing blue-green LEDs and lasers. A serious deterrent to its use, however, has been the difficulty in doping it p-type to a sufficient degree to permit good p–n junctions to be made. This kind of doping limitation, either n-type or p-type, is shared with most other wide-gap II–VI semiconductors, and has been the subject of study for many years; these studies have been reviewed by Neumark (1992) and Chadi (1994). Until recently the accepted explanation for all these materials was that the introduction of dopant atoms induces into the host material the simultaneous formation of native defects – that is, vacancies, interstitials, or antisites – that compensate the dopants (Mandel, Morehead, and Wagner 1964). These defects are not generally metastable, so they do not come within the purview of this book; however, the universality of this self-compensation explanation for stoichiometric ZnSe prepared at normal temperatures is now in question as the result of a calculation showing that the densities of native defects are about six orders of magnitude too small to account for the limitation of p-type doping (Laks et al. 1991). It has been pointed out, however, that departures from stoichiometry (which are common in these materials) can introduce many native defects that may interact with added dopants in ways that are difficult to predict (Chadi 1993). It has also been discovered that ZnSe prepared by atom-beam doping during molecular-beam epitaxy can be doped p-type up to densities of 3.4×10^{17} cm^{-3} with N atoms (Park et al. 1990b). Evidently, however, if hydrogen enters these systems it may passivate the N acceptors.

With conventional self-compensation by spontaneously created native defects now in doubt, there are several competing alternative explanations for doping limitations; each case needs a careful quantitative calculation to identify its magnitude. One of these explanations, related to metastable defects, is the possibility of either a substitutional or interstitial configuration for a dopant, corresponding to either shallow or deep levels for the carriers. Just as with the DX center in $Al_xGa_{1-x}As$ when $x > 0.22$, the ground state of some p-type dopants — including P and As — in ZnSe is not a substitutional configuration, so the dopant level is deep and holes are not free at normal temperatures (Chadi and Chang 1989a). It should be noted that the shallow-level metastable state of these dopants is at a considerably higher energy than the ground state, so it is not generally observed and these materials are not thought of as being metastable in the usual way. Nevertheless, the parallel with the DX center is otherwise complete. In this respect the problem of doping limitation and metastable defects are connected, just as in AlGaAs. For the case of N in ZnSe, its very small covalent radius seems to keep it in a fourfold, substitutional configuration with an effective-mass state for the hole that is weakly bound.

Among other sources of limitations to doping is a limited solubility of some dopants in a host. If the solubility limit is exceeded in an attempt to dope heavily, a second phase will form, creating precipitates or dislocations. Here, too, detailed quantitative calculations are required to tell if solubility effects are important, and they have recently been incorporated on an equal footing with shallow–deep transitions (Van de Walle et al. 1993). Although these solubility calculations have not yet included any relaxation of the dopant atoms away from the ideal lattice site, they have been able to explain the fact that ZnSe is difficult to dope p-type although ZnTe is very easy to dope p-type (Laks et al. 1993).

A different approach to the problem of deep vs. shallow levels for dopants was taken by Dow, Sankey, and Kasoowski (1991), who analyzed the absolute energies of foreign atoms relative to the host crystal energy bands. They showed that the valence-band maximum and conduction-band minimum for different host materials can be considerably different on an absolute energy scale, although the energy level of a foreign atom is more or less independent of the crystal bands. For example, E_V for ZnSe lies about 1 eV lower than E_V for ZnTe, although the impurity level is at the same absolute energy; thus the deep–shallow relationship for a single dopant atom can be substantially different in the two materials. Still to be included in this analysis are local structural relaxations.

Still another approach to this problem is that of Walukiewicz (1994) based on the properties of amphoteric native defects, which had been shown to control the stabilization of the Fermi energy in undoped GaAs, and is proposed to influence the doping efficiency by self-compensation in III–V and II–VI compounds. In this picture the formation energies of the amphoteric defects are dependent on the Fermi level in the manner developed by Baraff and Schluter (1986) and thus interact with doping.

These various explanations of doping limitations are still being debated. It appears that no one of them alone is capable of providing a universal solution of the problem, but the newest detailed calculations are providing the bases for eventual identification of which of the different contributions is dominant in each case.

3.3 Single-Crystal Silicon

Less metastable behavior of defects has been reported in Si than in the III–V compounds, probably because the smaller band gap requires that relevant energy barriers and energy differences for different gap states must be small. At low enough temperatures, however, the smaller energy differences become observable, and the same kind of behavior does occur in several cases. Annealing of these metastabilities occurs below room temperature, so there is less impact on ordinary device performance. Both the experimental evidence and theoretical interpretations for these defects have been reviewed recently by Watkins (1991). In addition, a number of defects have been produced in Si by radiation damage, which has been extensively studied, largely to elucidate the behavior of Si devices in space or in nuclear reactor environments. Radiation damage is itself a large field that we shall bypass here with one important exception.

That exception is V_{Si}, the vacancy in Si, which can be produced only by radiation damage at low temperatures; quenching from high temperatures has not succeeded, apparently because the vacancy is too mobile. Interest in V_{Si} follows from the unusual properties of its structure and electron levels, and from the theoretical advances that have occurred in explaining these properties. All of these have been reviewed by Watkins (1992), who is one of the pioneers in the study of the vacancy. The key feature of V_{Si} is its "softness"; the neighboring atoms can adopt a variety of configurations, depending on how many electrons occupy it, that is, the charge state of the center. One important configuration that does *not* occur is the retention of unperturbed lattice symmetry, leaving four dangling (unsatisfied) bonds around the vacancy; this configuration has a very high energy. Instead, the neighbors undergo Jahn–Teller distortions that reduce the symmetry of the center, and broken bonds reconstruct in various possible ways suggested in Figure 3.16 (Watkins 1992). The center seems capable of having as many as five states of charge (++, +, 0, –, ––)

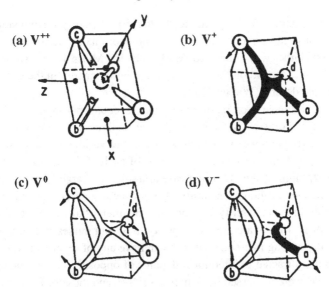

Figure 3.16 Various possible ways in which broken bonds around a Si vacancy can reconstruct depending on the charge state. *Source:* Watkins, in *Deep Centers in Semiconductors,* 2d Ed., ed by S. Pantelides (Gordon and Breach, Yverdon/New York, 1992), p. 177.

each with its own atomic configuration, and with energy barriers between them that produce metastable properties.

Theoretical analysis (Baraff et al. 1980) of this defect has led to the conclusion that the energies of the charge states of V_{Si} do not have the normal order in which more electrons correspond to higher energy, because it is found that reconfigurations result in a negative effective electron correlation energy (i.e., negative U; see Section 1.8), which brings the total energy of the relaxed single-positive state below all of the others. This conclusion has been tested since, and considerable experimental support has been described (Baraff et al. 1980).

A different class of metastable defects arises upon annealing of Si containing oxygen (which most Si does). These have come to be called *thermal donors* (TDs), and although they have been observed for many years, there is still no microscopic explanation for them (Wagner and Hage 1989). There are several of these TDs, some single donors and some double donors (TDDs). The properties of TDDs have been studied recently, with the conclusion that there are several of them with somewhat different energies. They are apparently all metastable at about 150 K, with optical excitation into the metastable state possible by band-gap light, and they all anneal below room temperature (Wagner and Hage 1989). Kinetics measurements of both formation and anneal at various temperatures indicate the need for thermal energy for transitions in both directions; thus a proposed configuration-coordinate diagram has barriers to transitions either way (Marenko, Markwich, and Murin 1985). These properties, however, depend on the position of the Fermi energy in the gap, hence on the sample doping (Watkins 1992).

Several cases have been reported of donor–acceptor pairing in Si. Binding between these atoms is by their Coulomb attraction, so they can form very simple systems having many stable states at the various possible spacings dictated by the lattice structure. The charge of each of the constituents influences the strength of the attraction, so variations in charge states make predictable changes in the pair stability. The most fully described examples of donor–acceptor metastability are for pairs of interstitial iron with acceptor atoms Al or B. Iron is a common impurity in Si and moves freely in a crystal, generally among interstitial sites. It can have any of several charge states, and is readily attracted to intentional dopants of opposite charge. The properties of Fe–Al were discussed by Scheffler (1989), and those of Fe–B by Kimerling (1988).

Substitutional nitrogen and oxygen in Si have been found to occupy off-center positions. That is, neither of their equilibrium positions is precisely at a substitutional site, and their neighbors relax into lower-symmetry configurations (Watkins 1991). Although metastable properties have not been reported for either of these, theoretical calculations suggest that, at least for oxygen, some states have energies deep within the valence band, so they are not observable. Thus all of the physics of metastability is present without the observable symptoms. Implanted hydrogen at low temperatures also has properties that have been interpreted as metastable (Holm, Neilsen, and Neilsen 1991).

3.4 Hydrogenated Polycrystalline Silicon

The photoinduced formation of metastable defects in hydrogenated polycrystalline silicon has been reported by Nickel, Jackson, and Johnson (1993). The properties of these defects are similar to those found in hydrogenated amorphous silicon (see Chapters 4 and 5): Electron spin resonance (ESR) measurements identify them as silicon dangling bonds (Johnson, Biegelsen, and Moyer 1982), hydrogenation decreases the density of dangling bonds (Kamina and Marcoux 1980), and hydrogen also decreases the concentration of weak Si–Si bonds at or near the grain boundaries (Jackson, Johnson, and Biegelson 1983). The principal difference between a-Si:H and poly-Si:H is that in poly-Si:H, strained Si–Si bonds, Si dangling bonds, and Si–H complexes are essentially confined to the two-dimensional grain boundaries. Another difference is that

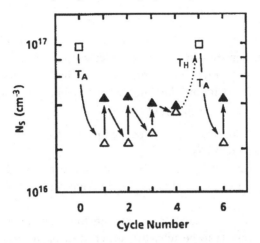

Figure 3.17 Spin density N_S of hydrogenated polycrystalline silicon films for several defect-formation and annealing cycles. ($\triangle$) Annealed state, ($\blacktriangle$) photodegraded state; degradation was carried out with 7 W/cm² white light for 15.5 h. ($\square$) Data obtained after hydrogenation at $T_H = 350\,°C$. Annealing was at 160 °C for 15 h. *Source:* Nickel, Jackson, and Johnson (1993).

a-Si:H displays only short-range order, whereas the grain boundaries in poly-Si:H show long-range order. A third difference is that the total number of metastable defects in poly-Si:H is generally smaller than that in a-Si:H since the defects are confined to the grain boundary regions.

Poly-Si films were deposited by low-pressure chemical vapor deposition at 625 °C to a thickness of 0.55 μm on quartz substrates (Nickel et al. 1993). The native oxide was removed with dilute HF to prevent a barrier to hydrogen incorporation. Hydrogenation was achieved by exposure to monatomic hydrogen from an optically isolated remote hydrogen plasma. Hydrogenation of poly-Si films for 7 h at 350 °C followed by annealing at 160 °C resulted in a metastable defect density of 2.2×10^{16} cm⁻³. Subsequent illumination with 7 W/cm² white light for 15.5 h at 300 K increased the defect density to 4.4×10^{16} cm⁻³. If this process of degradation and annealing is followed through several cycles, it is found that the minimum density N_{min} of metastable defects after annealing increases with cycle number, but the density of defects induced by the light N_s decreases with cycle number, as shown in Figure 3.17. Thermal annealing alone for up to 60 h at 160 °C did not cause an increase in N_{min}, indicating that the exposure to light between annealings was essential. If after four such cycles, the sample was reexposed to monatomic hydrogen at 350 °C for 60 min, the original annealed and light-degraded defect densities were restored, as in the initial sample conditions.

These results were interpreted by Nickel et al. (1993) as indicating that hydrogen is directly involved in the formation and annealing of defects, while at the same time recognizing that a detailed microscopic model cannot presently be given. They suggest that only hydrogen, dangling bonds, and strained Si–Si bonds are necessary for the photoinduced metastability, and advance the following sequence as a working hypothesis:

1. The initial hydrogenation of the poly-Si passivates Si dangling-bond defects, and also breaks weak Si–Si bonds.
2. Since the density of incorporated H greatly exceeds the density of dangling bonds, most of the H is located in regions not involving dangling bonds.
3. Annealing of the material without exposure to monatomic H causes weakly bound excess H to be released, which then passivates additional Si dangling bonds.
4. Illumination lowers the formation energy of dangling-bond defects to arrive at a new steady state with an increased dangling-bond density.
5. With repeated cycles, however, an increasing fraction of the H becomes located at more stable sites where it can no longer be released by illumination or annealing, and thus can no longer play the role described above for these two treatment conditions.
6. Reexposure to atomic H provides more weakly bound H to play its role in stabilizing and passivating dangling-bond defects.

It is proposed (Nickel et al. 1993) that a similar mechanism holds for both poly-Si:H and a-Si:H.

4

Hydrogenated Amorphous Silicon:
Properties of Defects

4.1 Amorphous Semiconductors and Their Defects

For the reader unfamiliar with amorphous semiconductors, we present a brief summary of some of the essential principles needed as a context for the treatment of metastable defects in hydrogenated amorphous Si (a-Si:H). For more detailed discussions of properties of amorphous solids the reader is referred to books by Elliott (1983) and Zallen (1983); for theory of tetrahedral semiconductors to that by Overhof and Thomas (1989); and for broad descriptions of a-Si:H to a four-part set edited by Pankove (1984), a two-volume set edited by Fritzsche (1989), and a book by Street (1991b). For additional historical background there is the landmark book of Mott and Davis (1979).

It is essential to recognize from the start that, although amorphous materials lack the long-range order of crystals, atoms in amorphous structures do not have random locations as occur in a gas. Rather, they retain the same short-range order that characterizes the local atomic relationships of a crystal of the same material; in this respect amorphous and liquid semiconductors have much in common. When some liquids are cooled sufficiently to solidify, they may form a glassy state, which is defined by such high viscosity as to provide structural rigidity (i.e., shear strength) although crystallization does not occur.

Most semiconductors, when cooled from the melt, form crystallites rather than glasses; but other more rapid cooling techniques can force them to become amorphous solids, although these are then limited to thin films. In fact, glasses are a subset of amorphous materials that can take this form by cooling from the melt. The amorphous state is not the state of lowest energy and, given enough time, these materials tend to *devitrify* or relax to a crystalline state that is structurally the ground state. Such devitrification, for example, was a serious problem for amorphous selenium used in some early Xerox machines. Hence the amorphous phase is metastable, and these relationships

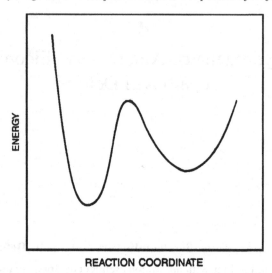

REACTION COORDINATE

Figure 4.1 Phase diagram for a two-phase system like a glassy material. The higher-energy metastable phase is the glassy state and the lower-energy ground state is crystalline.

may be represented by a diagram such as Figure 4.1, in which the abscissa is termed a *reaction coordinate* (among chemists) or an *order parameter* (among physicists). The similarity of this macroscopic description to configuration-coordinate diagrams used to describe the microscopic properties of localized defects occasionally leads to some confusion. Generally, the amorphous state of semiconductors is sufficiently enduring that, over a time of observation, it may be treated as structurally stable on a macroscopic distance scale. The electronic properties, which have much shorter characteristic times, may change many times while this structure is preserved. It is common, therefore, to refer to an *equilibrium* condition of the electronic system even though the underlying structure is not in its equilibrium state. Metastable defects, however, couple electronic and local structural elements, so their equilibration is necessarily more complicated and covers wide time scales.

The electrical and optical properties of amorphous semiconductors are often similar to those of the corresponding crystals (or liquids): There clearly are optical and electronic energy thresholds that resemble electronic energy bands and energy gaps. This fact is strong evidence for the earlier assertion that local bonding relations are maintained, and for the importance of bonds as a basis that is complementary to periodic potentials in establishing energetic relationships and electronic properties. On the other hand, the absence of translational symmetry in the material means that momentum is not a constant of electronic motion, and specific phonon-assisted transitions should not be observable. Reduced translational symmetry corresponds to spatially varying potentials, which can arise from structural nonperiodicity or from charged chemical impurities as in doped crystalline semiconductors (Ley 1984). The effect of structural disorder has been calculated theoretically in a-Si:H and found to be very strong, with a root-mean-square (rms) width of 0.2 elementary charges (Guttman, Ching, and Rath 1980).

Potential fluctuations, whatever their origin, create electronic states whose spatial average is used to describe the density of states; this average obscures the localized character of some of the states, which limits electronic transport through them at low temperatures. These states form the (nearly exponential) *band tails* that prevent a true gap from existing in the averaged density of states. In these tails, carrier mobility falls to low values at low energies, and a threshold in the energy spectrum of mobilities is formed, the *mobility edge*. Between the energies of the mobility edge for electrons in the conduction band and the edge for holes in the valence band is the *mobility gap,* which is what is generally meant by a gap in an amorphous semiconductor. It is close in energy to the observed thresholds for optical absorption, which depends more on the density of states and transition matrix elements than on transport.

Hydrogenated amorphous Si illustrates these points well: In electronically useful material there is a clear optical threshold of ≈ 1.75 eV (cf. 1.1 eV in crystalline Si), and the fundamental optical absorption edge is quite sharp, much more like that of a direct-gap material than the indirect-gap crystalline Si. These properties are responsible for its attractiveness as a material for solar-energy conversion, since a layer only ≈ 1 μm thick can effectively absorb most of the sunlight. Measured values of mobility gap are close to this value, perhaps slightly larger.

On the microscopic level, diffraction patterns of good a-Si:H show that nearly every Si atom is bonded to four others in the same tetrahedral configuration as in a crystal, and with a nearest-neighbor distance within 1 percent of the crystal value. Even the second neighbors have similar distances, although beyond that the similarities become lost. The principal distinction is that in the amorphous material there is about a 10° range of bond angles centered at the tetrahedral value of $\approx 109°$. The density of good a-Si:H is also within a few percent of that of the crystal. Stimulated by these observations, it has been shown that a complete amorphous network of tetrahedrally bonded atoms can be formed, retaining the short-range order except for small deviations in bond angles. This is called a *continuous random network* for the long-range randomness that results; it is regarded as the ideal for amorphous materials as a perfect crystal is for crystalline materials.

4.2 The Dangling-Bond Defect in a-Si:H

It is in this context that a native defect in an amorphous material can be defined. Rather than speak of a vacancy or interstitial as in a crystal, a native defect in a-Si is a departure from the tetrahedral coordination of some Si atom to four other Si atoms. It is found that all films of a-Si by itself contain many ($\approx 10^{19}$ cm^{-3}) such coordination defects, whose carrier-recombination properties and pinning of the Fermi energy make the material electronically useless; but the presence of about 10 percent hydrogen (usually from silane gas feedstock) greatly reduces the density of these defects and can make a-Si:H electronically useful (Chittick, Alexander, and Sterling 1969). This material can then be doped *n*- or *p*-type in familiar ways to make useful devices (Spear and Le-Comber 1976). It is the remaining defects – usually at densities of $\approx 10^{16}$ cm^{-3} – and the increases in their number induced by light (or other excitation) that are the subject of much study and analysis.

It is noteworthy that a few key properties of these defects appear to be shared in all the materials made in various laboratories and with a range of techniques. One of the most important is their electron spin resonance (ESR), which is found in every material at a *g* value of 2.0055, whether or not it is hydrogenated. The strength of this signal is taken as a measure of the bulk density of these defects (Taylor 1984), and since its discovery (Brodsky and Title 1969) its similarity to those on clean (111) surfaces of crystalline Si has identified it as an unterminated Si bond. Hence this dominating bulk defect in a-Si:H is denoted the *dangling bond* (DB). This identification has been supported by much evidence and correlation with other types of measurements, although, of course, ESR senses only those dangling bonds having a single electron in them, since those with either zero or two electrons are diamagnetic. This DB defect is illustrated in Figure 4.2 (Mott and Davis 1979). In the bulk of crystalline Si such a defect does not occur because a vacancy would require four DBs and have too high an energy; instead, bond reconstruction and symmetry lowering occur.

In current state-of-the-art undoped a-Si:H, the defect density is normally a little below 10^{16} cm^{-3}, but the density of DBs can be increased in a wide variety of ways: doping with the common dopants (phosphorus for donors or boron for acceptors) (Beyer, Mell, and Overhof 1977; Austin et al. 1979), quenching from elevated temperatures ($\approx 220\,°C$) (Street, Kakalios, and Hayes 1986), exposure to strong light (the Staebler–Wronski [1977] effect) or to electron beams (Schade 1984), the passage of forward currents in device structures (Staebler, Crandall, and Williams 1981), and formation of a space-charge region (Deane and Powell 1993b). Also, if a-Si:H is deposited onto substrates

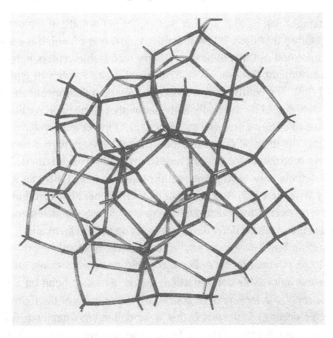

Figure 4.2 Pictorial representation of a continuous random network of tetrahedral bonds containing one dangling bond (unterminated bond at center of photo). *Source:* Mott and Davis, *Electronic Processes in Non-Crystalline Materials,* 2d Ed. (1979), by permission of Oxford University Press.

at temperatures well below the optimum ≈250 °C, an abnormally high density of defects arises (Street and Winer 1989). In most of these cases subsequent annealing at ≈200 °C can reduce these excess densities: The enduring increase in defect density induced by some nonequilibrium process, and the subsequent annealing away of the excess are the identifiers of their metastable character. The ground state of the center in a-Si:H seems electrically inert and is not seen directly in any measurement; only the metastable states are observable, so the total number of centers is always unknown. An important question on which disagreement continues is whether the *built-in* defects are distinct from those that can be introduced by subsequent excitation. Here, too, all are identified as dangling bonds by the same ESR signature, and many of their other properties are similar. Even more important is the disagreement on whether the defects have an origin that is inherent to this material or that is extrinsic (due to some physical or chemical abnormality). Many aspects of these stability questions were treated in the topical conferences on this subject (Stafford and Sabisky 1987; Stafford 1991).

It was thought earlier that the fact that some, but not all, of the defects anneal out required that there be two distinct kinds, one of which is stable; but it has been pointed out that annealing of any metastable centers must always leave some remnant fraction in their metastable state, even in equilibrium (Redfield 1988). The equilibrium fraction of centers in their metastable state is a quantity that should be thermally activated with a slope in an Arrhenius plot equal to the defect-formation energy. Moreover, if lower temperatures are used for annealing, the thermal relaxation rates over the energy barrier between the two states become much slower and exceed normal observation times. Thus the density of defects may have a nonequilibrium value even after an annealing (Smith and Wagner 1985). We shall not distinguish here between built-in and *excess* defects, except for quantitative purposes; hence the statement that the nature and origin of these defects are unknown applies to them all.

The dangling bond may have any of three states of charge (+, 0, −), corresponding to its occupation by either zero, one, or two electrons; all of these states have the metastable configuration, that is, a broken bond on a Si atom. The charge state is determined, of course, by the relation of the Fermi energy E_F (hence the doping) to the *levels* that were defined in Chapter 1, following Baraff, Kane, and Schluter (1980). These relations are expressed in CC-like diagrams in Figure 4.3, for which the coordinate Q is chosen to be the spacing of neighboring atoms along the direction of the dangling bond. For all three defect states the restoring force for motion along this direction should be weaker than it is for the ground state with its fourfold coordination. For this reason, the defect curves are drawn broader than that of the ground state; although they are drawn as parabolas, the low symmetry of the defect means that in actuality these curves should be asymmetrical about their minima, causing anharmonic effects that are undoubtedly important. The equilibrium spacings for the three defect states in Figure 4.3 (the values of Q at the minimum energies) are taken to be close to each other, but all very different from the spacing in the ground state.

According to Adler (1984), each of these three charge states of the defect has a different electron hybridization. In Adler's picture the neutral charge state has three of the same sp^3 hybrid bonds as for tetrahedral bonding, even though the fourth bond is broken. The positive charge state, in which the dangling bond contains no electrons, is associated with rehybridized bonds that become sp^2, and the negative state is associated with p^3 bonding. With each variation in charge state there is a comparatively small change in the position of the central atom (and hence in the value of Q). It is also likely that the restoring forces are different for the three states, an effect that is shown pictorially by giving the three defect parabolas arbitrarily different shapes.

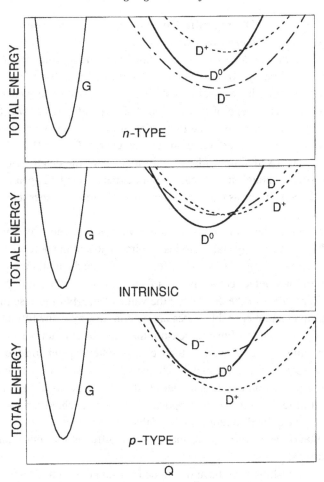

Figure 4.3 Schematic configuration-coordinate-like diagrams for the dangling-bond center in a-Si:H with three possible metastable states corresponding to three states of charge. In this example, the configuration coordinate Q may be taken as the interatomic distance along the direction of the dangling bond. The upper parts of these curves are not known. The illustrated differences caused by doping are explained in the text.

When material is n-type, E_F is high in the gap and all the defect states tend to be occupied by electrons, so that the D^- state of the defect has the lowest energy, as shown in the top panel of Figure 4.3. For p-type material the D^+ state has the lowest energy, and for intrinsic material the D^0 state has the lowest energy, as shown in the other panels. The total energies of the D^- and D^+ states shift with E_F to create these relations because E_F contributes to their energy as was shown in Eq. (1.1). The energies of the ground state G and neutral state of the defect D^0 are unaffected by changes in E_F because they do not involve exchange of an electron with the reservoir of electrons formed by the rest of the system. The *levels* $E(-/0)$ and $E(0/+)$ are the energies that E_F must take for the minima of two curves, D^-, D^0 and D^0, D^+, to become equal, so that their occupancies have equal probabilities; the values of these energy levels are discussed at the end of this section. This type of diagram was first introduced for the elucidation of the properties of the vacancy in crystalline Si, inferring there a negative effective correlation energy U (Baraff et al. 1980). Branz (1989) invoked similar ideas in his analysis of the dangling-bond defects in a-Si:H, although in comparing with experiments he appears to misinterpret DLTS energies as equilibrium energy differences rather than activation energies.

All three defect states of Figure 4.3 are always present, no matter what the doping; the higher-energy states are just excited states of the defect. Also, the defect-formation energy (the energy difference between the minimum of G and the minimum of a defect curve) is not a fixed, single quantity; it depends on both the charge state of the defect and the value of E_F. (For the case in which both the ground and metastable states are neutral, however, the formation energy has a single value.) For this reason, the equilibrium number of defects at a given temperature, as well as their charge state, depends on the doping through E_F; specifically, there may be charged defects even in intrinsic material, although they should be less numerous than neutrals. As always in cases of large lattice relaxation, however, optical excitation thresholds and activation energies may be quite different from the defect-formation energies. In fact, the interpretation of densities of states may be quite different for optical and thermal transitions.

These relationships have been developed for homogeneous materials with discrete energy levels, but they are also applicable to defect states in materials with inhomogeneities (Branz and Silver 1990). In such materials the inhomogeneities produce a distribution in the values of defect parameters, and these relations must be applied to each local condition that determines parameter values. Distributions of values are expected in amorphous materials, and there is much evidence for their presence; they have overriding importance in the *defect pool* models discussed in Section 5.6.

Because of the complexity of Figure 4.3, a simplified representation is generally used that may be satisfactory for undoped material. Shown as Figure 4.4, this presents only one metastable energy well of depth E_2, which is the annealing activation energy. The barrier seen by the inactive ground state is E_1, which is not readily measured. However, $\Delta E \equiv E_1 - E_2$ has been inferred from high-temperature measurements of the equilibrium density of defects,

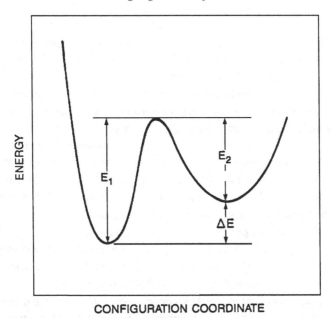

CONFIGURATION COORDINATE

Figure 4.4 Simplified configuration-coordinate diagram that is generally used for the dangling-bond defect in undoped a-Si:H.

leading to the value $\Delta E \approx 0.2$ eV. It must be cautioned that Figure 4.4 obscures the fact that the defect-formation energy ΔE is not a simple constant. Moreover, when one incorporates a distribution of values for the defect energies into the analysis, there are two qualifications necessary in the interpretations of these energies:

1. The *defect-pool* treatment summarized in Section 5.6 leads to the conclusion that the activation energy $\Delta E \approx 0.2$ eV from quenching measurements is much lower than the actual average formation energy, which Schumm (1994) says is 0.9 eV.
2. It has also been pointed out that, for the observed stretched-exponential forms of annealing transients, the experimental activation energy does not equal E_2 when the stretch parameter β is linearly temperature-dependent, as has been reported, but rather E_2/β (Benatar, Redfield, and Bube 1993). Since annealing activation energies are typically ≈ 1 eV and β is commonly ≈ 0.5 near room temperature, this implies that E_2 is only ≈ 0.5 eV, much lower than generally recognized.

Although these centers show no activity when in their ground state, they are remarkably active in their metastable state. In addition to their ESR signal when neutral, these gap states act as carrier recombination centers and contribute to extrinsic optical absorption and photoconductivity. Quantitative interpretations of experimental data to infer values for these energies frequently approximate by discrete values (see Section 5.7) the likely distributions in values caused by the nonidentical environments of the centers in an amorphous structure. The many efforts that have been made to determine their density of electronic states as a function of energy in the gap have produced a variety of results, but there is some consensus. Most investigations indicate that for undoped material the value of $E(-/0)$ at room temperature is close to 0.90 eV above the valence-band mobility edge. Initial work (summarized in Cody 1984) using optical absorption indicated that $E(-/0)$ was slightly larger than 0.90 eV, and Stutzmann and Jackson (1987) concluded from a study of variously doped a-Si:H that $E(-/0) = 0.95$ eV. Using a model of photoconductivity applied to the data of LeComber and Spear (1986) on the variation of photoconductivity with E_F, Bube and Redfield (1989b) concluded (see Section 5.7) that the best fit was obtained for $E(-/0) = 0.90 \pm 0.02$ eV.

From an analysis of extensive dark-conductivity data giving the dependence of the Fermi energy on the defect density and temperature for a typical undoped a-Si:H sample, Bube, Benatar, and Redfield (1994) concluded (see Section 4.6) that the data required at least two different sets of levels with slightly higher room-temperature values: $E_1(-/0) = 1.06$ eV and $E_2(-/0) = 0.97$ eV, under the assumption that $U_1 = U_2 = 0.20$ eV. The value of $E(0/+)$ is less well known since it is in the lower half of the energy gap and most material is n-type or intrinsic. Values between 0.20 and 0.45 eV have been inferred for the effective correlation energy $U = [E(-/0) - E(0/+)]$.

4.3 Defects and Doping in a-Si:H

Although the emphasis of this book is on metastable, light-induced effects of defects, the case of light-induced dangling bonds in a-Si:H cannot be fully treated without reference to studies of the built-in defects of the same kind – that is, those DBs present in as-deposited films, whether before light exposure or after annealing away of light-induced defects. The densities of such defects have often been referred to as the *equilibrium* densities, and there are many theories to describe their properties; the values observed at room temperature, however, could simply represent frozen-in defects in a nonequilibrium state that has not relaxed long enough, so their interpretation requires care.

It has been long known that simply doping the films either n-type (e.g., with phosphorus) or p-type (with boron) quenches the photoluminescence, suggesting that there is an increasing density of carrier-recombination centers

(Tsang and Street 1978; Austin et al. 1979; Fischer et al. 1980). These effects were studied systematically by combining photoluminescence, ESR, and light-induced ESR (LESR) on a series of samples that were doped n-type or p-type at various levels or compensated with both B and P (Street, Biegelsen, and Knights 1981). The clear findings are that the density of dangling-bond defects increases with n- or p-type doping but decreases with compensation. These conclusions were confirmed using photothermal deflection spectroscopy (PDS), under the assumption that the subgap optical absorption coefficient is proportional to the density of deep-level defects – an assumption confirmed by comparing the results of optical absorption and ESR and LESR (Jackson and Amer 1982; Amer and Jackson 1984). These effects of doping are summarized in Figure 4.5.

In an extension of such work, Stutzmann and Jackson (1987) combined electrical transport measurements with ESR and PDS on sets of n- and p-type samples, observing particularly the differences in the densities of paramagnetic (neutral) defects for different doping-controlled values of the Fermi level E_F. Since low values of E_F cause the defects to become positive and high values cause them to become negative, both of which are diamagnetic, it was possible to infer the energies of the transitions levels $E(0/+)$ and $E(-/0)$, which they reported as separated by the effective correlation energy $U = 0.2$ eV. This was the first attempt to incorporate the changes in total defect density (obtained by PDS, described in Section 4.5) due to doping into such an analysis.

To study light-induced defects, the PDS technique was similarly applied in samples of various dopings with the finding that increased doping (either n- or p-type) results in higher densities of light-induced defects (Skumanich, Amer, and Jackson 1985). Thus, as doping changes, there is a correlation between the resulting density of built-in defects and that of light-induced defects; as shown in Figure 4.6, their relation is linear, except for the lowest densities. There have been many other reports since showing such a correlation between the densities of light-induced and built-in defects, although there are exceptions. For compensated materials in Figure 4.6 there is a reduction in the density of light-induced defects, and in fact the density of these defects in closely compensated material is significantly lower than in the best undoped material. This fact must be considered in theories of these defects; it is mentioned later in the discussion of models (Section 5.6).

Because of these close similarities between defects induced by light and by doping, virtually all attempts to explain light-induced defects have had to include these others in some way. This sometimes amounts to equating equilibrium and nonequilibrium behavior, a practice that requires care and justification in all cases.

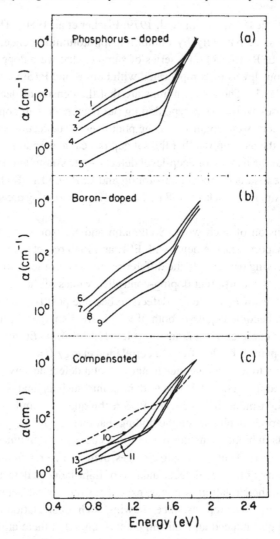

Figure 4.5 Optical absorption spectra using photothermal deflection spectroscopy on samples doped either (a) *n*-type, (b) *p*-type, or (c) compensated. (a) PH_3 doping concentrations of (1) 10^{-2}, (2) 10^{-3}, (3) 3×10^{-4}, (4) 10^{-5}, and (5) 10^{-6}. (b) B_2H_6 doping concentrations of (6) 10^{-3}, (7) 3×10^{-4}, (8) 10^{-4}, and (9) 10^{-5}. (c) All samples of compensated material have 10^{-3} PH_3; the B_2H_6 concentrations are (10) 2×10^{-4}, (11) 4×10^{-4}, (12) 2×10^{-3}, and (13) 4×10^{-3}. All doping concentrations refer to gas-phase concentration. Substrate temperature is 230 °C and the rf deposition power is 2 W. *Source:* Amer and Jackson (1984).

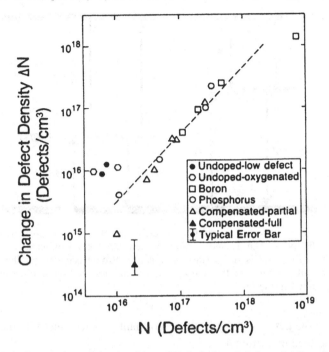

Figure 4.6 Correlation of the densities of light-induced defects (ordinate) ΔN with the density of built-in defects N present in the starting material. *Source:* Skumanich, Amer, and Jackson (1985).

4.4 Description of Dynamic Behavior in Terms of Energy Levels

Modeling of a variety of electronic phenomena involving the metastable DB defects in a-Si:H has been carried out using a discrete-energy-level model of a single type of multivalent defect. Such a discrete-level model is certainly a simplification, and in a number of cases more complex distributions – for example, discrete levels corresponding to several different types of multivalent defects, or a Gaussian distribution of level density with energy – have been assumed with good agreement with experiment, as described in later sections. Also the defect-pool model described in Section 5.6 has a very different description. From the successes achieved with the discrete-level model for a multivalent defect, however, it appears likely that the major effects of the defects can be described, at least approximately, in this way. Such a model is especially useful in describing the photoconductivity effects in the material (see Section 5.7), particularly since photoconductivity is one of the methods used

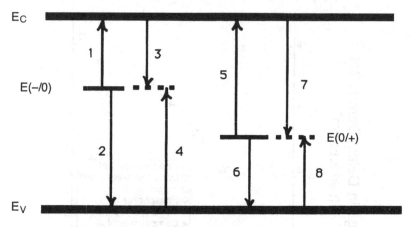

Figure 4.7 Energy level diagram for a multivalent defect with possible charge states of −, 0, and +. The dashed portion of each defect level indicates the condition in which the level is in its state of lower electronic occupancy. The arrows represent possible optical or thermal excitation energies, and recombination energies as described in the text.

to measure the defect density itself (in constant-photoconductivity measurement, described in Section 4.5).

The energy levels for dangling-bond defects in a-Si:H are are similar to the energy levels for Cu impurity in *n*-type crystalline ZnSe, for example, with which they can be compared. The DB defect can be described as existing in three states:

1. a diamagnetic, negatively charged state occupied by two electrons (like Cu^{-1} in ZnSe),
2. a paramagnetic, neutral state occupied by one electron (like Cu^0 in ZnSe), and
3. a diamagnetic, positively charged state with no electrons (like Cu^+ in ZnSe).

The actual energy values of the two energy levels $E(-/0)$ and $E(0/+)$ in a-Si:H are determined as described in Section 4.2 by the values of the Fermi energy at which the minima of the corresponding CC curves are equal.

In order to use this type of model in dynamic processes involving optical and thermal excitation, as well as capture of free carriers, an energy-level diagram such as that given in Figure 4.7 is conventional. A total of sixteen different transitions can formally be defined as necessary in the general case:

1. An optical excitation with energy $E(1) = [E_C - E(-/0)]_{opt}$ that raises an electron from the negatively charged defect to the conduction band by

absorption of a photon, or a thermal excitation with energy $E'(1) = [E_C - E(-/0)]_{th} \leq E(1)$ that raises an electron from the negatively charged defect to the conduction band by absorption of thermal energy.

2. The released energy $E(2) = E(-/0)$, either in the form of a photon or phonons, when an electron at the $E(-/0)$ level recombines with a hole in the valence band.

3. The released energy, either in the form of a photon or phonons, with capture of an electron from the conduction band by the neutral defect with energy $E(3) = [E_C - E(-/0) - E^*(-/0)]$, where $E^*(-/0)$ is the lattice-relaxation energy involved when the charge of the defect changes from negative to neutral.

4. An optical excitation with energy $E(4) = [E(-/0) + E^*(-/0)]_{opt}$ that raises an electron from the valence band to the neutral defect by absorption of a photon, or a thermal excitation with energy $E'(4) = [E(-/0) + E^*(-/0)]_{th} \leq E(4)$ that raises an electron from the valence band to the neutral defect by absorption of thermal energy.

In cases of strongly localized centers with large lattice relaxations (e.g., the DX center), there may be significant differences between the thermal and optical transition energies, although they are not explicity shown in Figure 4.7. In many practical cases especially in covalent semiconductors, however, the assumption is often made that for first-order purposes $E(1) \approx E'(1) \approx E(3)$ and $E(2) \approx E(4) \approx E'(4)$. Experimental evidence, such as the small value of the Stokes shift in luminescence, indicates that this is a reasonable approximation in a-Si:H. The other transitions shown in Figure 4.7 involving the (0/+) state of the defect can be described analogously. Applications of this model to dark-conductivity data are described in Section 4.6, and those to photoconductivity data in Section 5.7.

4.5 Techniques for the Measurement of Defect Density and Their Limitations

Several different techniques have been used to obtain information about the density and other properties of metastable defects in a-Si:H (e.g., Street 1991b, pp. 104–30). Since in high-quality, undoped material, the defect density is in the 10^{15}–10^{16} cm^{-3} range, the optical absorption coefficient is of the order of 1 cm^{-1} and is not readily measurable by direct optical transmission techniques in typical films with thickness of 1 μm.

Since the initial suggestion of Grimmeiss and Ledebo (1975), considerable use has been made of the technique known as *constant-photoconductivity measurement* (CPM) (Moddel, Anderson, and Paul 1980; Vanecek et al. 1981; Gunes et al. 1991). Conventional photoconductivity spectra use a constant value of light flux L (photons per unit area per unit time), and the resulting values of the photoconductivity $\Delta\sigma$ vary because of the spectral sensitivity of the absorption coefficient α and the carrier lifetime τ. In contrast, CPM uses varying values of L to maintain a fixed value of $\Delta\sigma$ as the spectrum is scanned, so that the free-carrier densities (and hence the quasi-Fermi levels) stay fixed.

CPM is one of three techniques frequently used to measure the density of defects in a-Si:H. The other techniques that have received wide use are photothermal deflection spectroscopy (Jackson et al. 1981), and electron spin resonance (e.g., Street 1991b, p. 104ff). Each method has its intrinsic limitations in providing an accurate value of the total defect density in a-Si:H. PDS is sensitive to the properties of the material's surface, where defect densities can be appreciably larger than in the bulk. Measurements on films with a number of different thicknesses are required to sort out bulk and surface effects (Shimizu et al. 1989). ESR is also surface-sensitive, and involves only those defects in a-Si:H that exist in the neutral state and hence have an unpaired spin; if a distribution of defects is present, only a fraction exist in this state, depending on the specific distribution and the Fermi energy. Additional information can be obtained by using bias lights of various intensities in connection with the ESR measurements. CPM has the advantage of being both surface-insensitive and able to measure charged and neutral defect densities, but may depend on the Fermi-level position and other details of the defect distribution.

Constant Photoconductivity Measurements (CPM)

Several investigations have been directed at determining the appropriate interpretation of CPM measurements under a variety of possible conditions (Marshall et al. 1991; Wyrsch et al. 1991; Liu et al. 1992; Zhong and Cohen 1992; Hattori et al. 1993; Platz, Brueggemann, and Bauer 1993; Siebke and Stiebig 1994). CPM makes use of the fact that, under appropriate circumstances, the photoconductivity measured for photoexcitation in the extrinsic range (i.e., photoexcitation of electrons from defects to the conduction states) is proportional to the absorption coefficient α for excitation from these states, and hence to the defect density itself. In its simplest form the photoconductivity $\Delta\sigma$ can be expressed as

$$\Delta\sigma = \Delta n\, q\mu \tag{4.1}$$

where Δn is the additional photoexcited electron density and μ is the electron mobility. Since in steady state $\Delta n = G\tau$, where G is the photoexcitation rate

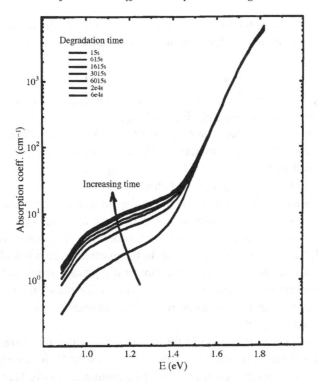

Figure 4.8 CPM-determined absorption spectra as a function of degradation time at 360 K under 1 W/cm² illumination ($G = 2 \times 10^{22}$ cm⁻³ s⁻¹), showing saturation of absorption (hence of defect density) for long times of degradation.

and τ is the electron lifetime, and since $G = \alpha L$, Eq. (4.1) can be rewritten (assuming unity quantum efficiency) as

$$\Delta\sigma = \alpha L \tau q\mu \qquad (4.2a)$$

or

$$\alpha = (1/L\tau)\,(\Delta\sigma/q\mu) \qquad (4.2b)$$

Absorption $\alpha(h\nu)$ is derived from CPM by measuring the photon flux L necessary to keep the photoconductivity constant over the spectrum of photon energies covered by the experiment. All of the parameters on the right side of Eq. (4.2b) except for L are assumed to remain constant as the photon energy is varied. Thus the result of a CPM experiment is a $1/L(h\nu)$ spectrum that should be proportional to $\alpha(h\nu)$. A sample set of absorption curves determined by CPM for different optical degradation times is given in Figure 4.8. For extrinsic excitation from the defect level to the conduction band, $\alpha_{\text{ext}} = S_{\text{opt}}N_{\text{I}}$, where N_{I} is the density of electron-occupied defects available for photoexcitation, and S_{opt} is their optical cross section. If $N_I = A'N_{CPM}$, where N_{CPM} is the total density of imperfections measured by CPM, Eq. (4.2) can be written as

$$N_{CPM} = (1/L\tau)/(\Delta\sigma/A'S_{opt}\, q\mu) \qquad (4.3)$$

Thus N_{CPM} is proportional to $1/L$, and the values of L needed to maintain $\Delta\sigma$ constant may be used to give changes in N_{CPM}.

This inverse relation between N_{CPM} and L does not require a linear dependence of $\Delta\sigma$ on G. In the common case in which $\Delta\sigma \propto G^{\gamma}$, where γ is some number less than 1, $\Delta\sigma \propto L^{\gamma}$, $\tau \propto L^{\gamma-1}$, and N_{CPM} is still proportional to $1/L$, provided that γ is not a function of photon energy.

Experimentally $1/L$ is scaled to α by making an independent measurement (by standard transmission methods) of α_{int} in the intrinsic region (e.g., at 1.7 eV). With α_{int}, $\Delta\sigma$, and L known, this procedure determines the value of $\mu\tau$ at 1.7 eV. By assuming that $1/L$ in the extrinsic region has also now been correctly scaled to $\alpha_{ext,}$, we are assuming that $\mu\tau$ as determined from the intrinsic portion of the spectrum is the same as for the extrinsic, defect-controlled portion of the spectrum; that is, we are assuming that $\mu\tau$ is not a function of photon energy.

Additional complications occur for conditions in which the Fermi energy lies near the defect level (becoming more severe as the Fermi energy drops below the level), which result in the CPM-determined density being lower than the actual defect density. This is a common occurrence in undoped, high-resistivity a-Si:H samples, and significant changes in the relative positions of the Fermi level and the defect level can occur during the process of optical degradation or thermal annealing. These phenomena have been explored (Bube et al. 1992) by considering several typical models, including the effects of capture coefficients, defect density, a Gaussian distribution of defect levels, and the possibility of photoexcitation to empty defects from the valence band.

Several experimental concerns that must be considered in developing an overview of the CPM technique involve:

1. the smoothing of the experimental data involving interference fringes for measurements on thin films;
2. a comparison of a.c. and d.c. photoconductivity measurements to test for time-constant effects as a function of photon energy; and
3. choosing how to relate the data to a specific defect density.

Because measurements on a-Si:H involve thin films, the CPM data show characteristic interference fringes. To use these curves to obtain defect densities, it is necessary to correct for the interference effects. An apparently effective method is to use Fourier smoothing (Wiedeman, Bennett, and Newton 1987;

Press et al. 1989) with a window width equal to the interference-fringe spacing.

The possibility of the existence of time-constant effects varying with photon energy is always present with the use of an a.c. CPM method if the photoconductivity time constant is longer than the reciprocal of the a.c. chopping rate, which leads to errors in the CPM-determined defect densities. One way to avoid this problem is to use d.c. CPM instead of a.c. CPM. A limitation of this approach is that the measured photocurrent must be greater than the dark current under the temperature and light-intensity conditions of measurement.

A reliable method is needed to derive values of defect density from the measured CPM data. For a-Si:H films it is common to use a value of 750 cm^{-1} for the absorption coefficient at 1.7 eV, obtained from detailed modeling of transmission data on a single sample (Bube et al. 1992). The next step is to correlate the absorption coefficient with the defect density. A number of proposals have been made; for instance, the defect density is given by 2×10^{16} cm^{-3} times the integrated subgap absorption, or 2×10^{16} cm^{-3} times the absorption at 1.2 eV (Wyrsch et al. 1991). A good correlation exists between the absorption at various subgap energies and the integrated subgap absorption.

Necessary conditions for the CPM density, determined by normalization of CPM spectra in the intrinsic range, to be accurate can be summarized as follows:

1. The free-electron lifetime τ must not be a function of photon energy.
2. The power dependence γ of photoconductivity must not be a function of photon energy.
3. For a.c. CPM measurements, the response time τ_0 must not be a function of photon energy. The use of d.c. CPM measurements can help overcome this difficulty.
4. The equilibrium Fermi level must lie at least 0.1 eV above the level of the defect being measured. If this is not the case, then it is possible to correct the CPM-measured densities by a knowledge also of the location of the equilibrium Fermi level relative to the defect energy level for each CPM-measurement condition (Bube et al. 1992). This correction requires a knowledge of the magnitude of the capture coefficients for the various relevant processes in the CPM measurement. If the Fermi level lies below the energy level of the defect being measured, correction also requires a knowledge of the magnitude of the optical cross section for excitation of electrons from the valence band to the empty level. If a Gaussian distribution of identical levels exists, the correction factor may be reduced for Fermi energies more than 0.05 eV below the maximum of the distribution.

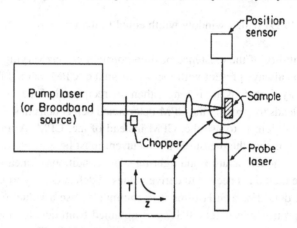

Figure 4.9 The experimental arrangement for the photothermal deflection spectroscopy method of defect density measurement. *Source:* Amer and Jackson (1984).

5. Only one type of defect must dominate. If there are a variety of different defects, with different energy levels, capture coefficients, and densities, it is highly likely that a CPM measurement will not yield the correct total defect density.

Overall the results emphasize the high importance of simultaneous measurements of CPM density, equilibrium Fermi energy, and photoconductivity at all stages of optical degradation and thermal annealing in order to allow an informed and accurate assessment of the magnitude and variation of the actual defect density and related effects in a-Si:H.

Photothermal Deflection Spectroscopy

The method of photothermal deflection spectroscopy (PDS) (Boccara, Fournier, and Badoz 1980; Boccara et al. 1980; Jackson and Amer 1981a,b; Jackson et al. 1981; Amer and Jackson 1984) utilizes the fact that absorption of light causes an increase in the temperature of the material. Figure 4.9 shows a typical experimental arrangement. The sample is immersed in a suitable liquid. When a chopped pump light is absorbed by the material, periodic heating occurs in the liquid near the surface of the sample, which then gives rise to a corresponding periodic variation in the index of refraction of the liquid near the surface. A gradient in refractive index, produced by the temperature gradient in the fluid, is developed near the sample surface. A weak laser beam is used to probe this gradient in refractive index by being directed in grazing incidence to the sample surface. A deflection of the laser beam caused by the gradient in refractive index is then detected and interpreted as a measure of the heating, and hence of the optical absorption. When the photon energy of the pump beam is varied, the corresponding magnitude of the deflection of the laser beam is a measure of the optical absorption, and hence of the defect density responsible for the absorption. The actual optical absorption can be obtained as the result of a modeling of the PDS effects (Jackson and Amer 1981a,b; Amer and Jackson 1984).

As stated above, the magnitude of the PDS signal is sensitive to the density of defects at the surface, which is usually larger than in the bulk. In order to use the PDS technique to measure bulk defect densities accurately, therefore, it is necessary to measure a series of samples with different film thicknesses.

Magnetic Resonance

Magnetic resonance in its various forms has been extensively applied to the study of defects in a-Si:H and has provided a wide range of information. An excellent review of the earlier work in both nuclear and electron spin resonance has been given by Taylor (1984); Morigaki (1984) reviewed optically detected magnetic resonance and electron-nuclear double resonance in the same volume. A more recent overview of ESR and light-induced ESR (LESR) by Tanaka's (1991) group showed how pulsed ESR can exhibit small hyperfine interactions and nuclear quadrupole interactions using the method of *electron-spin-echo-envelope modulation*. There are, however, a range of interpretations of resonance results (as with others in a-Si:H), an example of which is the still more recent work of Hari, Taylor, and Street (1993, 1994), whose interpretation of hydrogen placement and motion disagrees with that of Tanaka.

Following the early ESR identification of the dominant paramagnetic defect in unhydrogenated a-Si by its gyromagnetic ratio g value of 2.0055 (Brodsky and Title 1969) as a dangling Si bond, it was found that its density is greatly reduced in a-Si:H, typically down to values around 10^{15} cm^{-3} in good undoped material. Many measurements then showed that in undoped a-Si:H the *only* ESR line found in equilibrium is this one. Doping of a-Si:H produces two effects seen with ESR: The strength of the $g = 2.0055$ line (and hence the density of dangling bonds) increases linearly for either n-type or p-type doping (Dersch, Stuke, and Beichler 1981a), and two additional resonance lines appear, one having $g = 2.0043$ for n-type doping, and one with $g = 2.01$ for p-type doping (Dersch, Stuke, and Beichler 1981b).

Photoexcitation of undoped a-Si:H also produces two distinguishable effects:

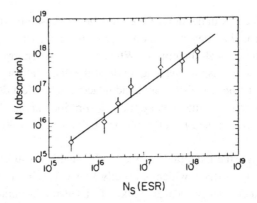

Figure 4.10 Comparison of the defect density measured by photodeflection spectroscopy N and electron spin resonance N_s in undoped a-Si:H. The defect density is varied by varying the deposition conditions. *Source:* Amer and Jackson (1984).

1. Weak light (LESR) introduces the same two new resonance lines as caused by doping; these quickly die out after the light is shut off (Dersch, Stuke, and Beichler 1981c). These other lines are generally interpreted as representing electrons in conduction-band-tail states and holes in valence-band tails, respectively. Nevertheless, Tanaka (1991) showed evidence for an alternative interpretation that these other two lines represent different kinds of singly occupied dangling-bond states or DB-like localized states.
2. Upon exposure to strong light the predominant effect in undoped a-Si:H is a metastable increase in the density of centers with the characteristic value $g = 2.0055$.

Later studies compared the densities measured by ESR and PDS, and found good correlation as shown in Figure 4.10 (Amer and Jackson 1984) – a correlation that is significant because PDS can sense all the defects, whereas ESR senses only those that are paramagnetic. The resulting inference that most of the defects in undoped material are neutral (paramagnetic) has important consequences for microscopic models. In a more comprehensive study, Stutzmann and Jackson (1987) measured both the PDS and ESR densities along with the conductivity on a series of samples with different doping. The values of the Fermi energy were inferred for each of these samples from the conductivity measurements, thus providing comparison of the total defect density with the neutral density as a function of E_F. The variations they found, shown in Figure 4.11, are attributed to the changing occupation of the defect states with changing E_F, which in turn permits inference of the values of the energy levels and the effective correlation energy, which they deduced to be 0.2 eV.

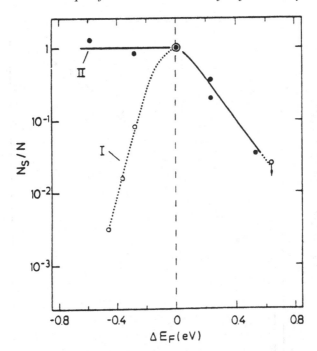

Figure 4.11 Dependence of the ESR spin density N_s as a fraction of the total defect density (measured by PDS) N on Fermi energy in a set of variously doped samples. The portion labeled II is for poor material, and is considered unimportant. *Source:* Reprinted from *Solid State Communications* 62, 153, Stutzmann and Jackson, "Occupancy of Dangling Bond Defects in Doped Hydrogenated Amorphous Silicon," © 1987, with kind permission from Elsevier Science Ltd., The Boulevard, Langford Lane, Kidlington OX5 1GB, UK.

Again appeared the finding that for intrinsic material (for which $\Delta E_F \approx 0$ in Fig. 4.11), essentially all the defects are neutral. Other studies demonstrated good agreement of the values of defect density in samples with a range of dopings when measured by LESR, luminescence, photoconductivity, and optical absorption (PDS), as shown in Figure 4.12 (Street 1984).

The kinetics of changes in ESR spin density were studied early by Stuke's group (Dersch et al. 1981a), who reported the time and temperature dependence of the annealing of the excess density, which agreed with the behavior of metastable defects observed by conductivity. Extensive series of measurements of the kinetics of defect generation (discussed in Section 5.1) were later made by two groups using ESR and given quite different interpretations. Those of Taylor's group (Lee, Ohlsen, and Taylor 1985), shown in Figure 4.13, covered temperatures of 77–500 K and have a temperature-dependent sat-

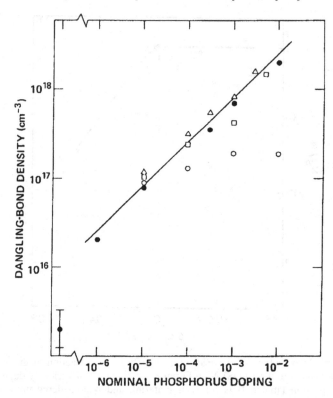

Figure 4.12 Comparisons of the defect density measured by ESR (O), lumines-cence (△), PDS (●), and photoconductivity (▢) for a number of phosphorus-doping levels. *Source:* Street (1984).

uration in the defect density at all temperatures. Stutzmann, Jackson, and Tsai (1985), whose main results are shown in Figure 4.14, reported no saturation below ≈370 K and transient responses of a very different character, which were described over a limited range as having (time)$^{1/3}$ dependence. The analyses and controversies that grew out of kinetics measurements are summarized in Chapter 5. Subsequently, saturation at essentially all temperatures has been con-firmed, and the time dependences have been replaced by stretched exponentials.

Nuclear magnetic resonance methods have also been widely applied to a-Si:H since NMR can serve as a detailed local probe of bonding configura-tions. It is particularly useful for studying the properties of hydrogen in a-Si:H because it is inherently sensitive to H atoms and because there are large quan-tities of H present. For example, it has been shown that H occupies two types of sites (in addition to small amounts of molecular H_2): one that is dilute and probably represents Si-H bonds, and another that is clustered, perhaps on in-ternal voids. The properties of H in a-Si:H are interesting and varied, and there

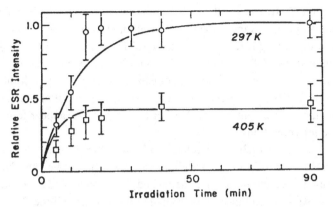

Figure 4.13 ESR defect density (points) as a function of time of exposure to white light at two temperatures. The curves are fits by a proposed model. *Source:* Lee et al. (1985).

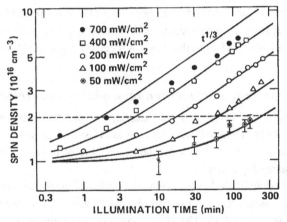

Figure 4.14 ESR spin density (points) as a function of time of exposure to several different intensities of white light (measured at room temperature). The curves are fits by a proposed model. *Source:* Stutzmann, Jackson, and Tsai (1985).

have, in fact, been many proposals (some discussed in Section 5.6) that directly involve H in the defect-state transitions. When deuterium is substituted for H in these films, additional information is provided by the nuclear quadrupole interaction, which is very sensitive to local bonding. Much of this has been summarized by Taylor (1984).

Both ESR and NMR have contributed in the continuing debate over a possible role of H in the dangling-bond defects (discussed in Section 5.6). In one study using deuteron magnetic resonance (DMR), Norberg et al. (1992) found a correlation over five orders of magnitude between the spin-lattice relaxation times and the photovoltaic quality of a range of materials. From photoinduced changes in the DMR spectra, they also concluded that "certainly the microvoid hydrogens and perhaps the [others] are decoupled from the locations of any light-induced paramagnetic electronic defects." This agrees with the resonance-based conclusions of Tanaka (1991) that both H and D avoid the immediate vicinity of dangling bonds. These conclusions are consistent with the argument that both DBs and H tend to reduce strain, and would thus not generally occur together; but recent results (Hari et al. 1993, 1994) appear to disagree with these interpretations.

Another area in which NMR relates to light-induced metastable changes was reported by McCarthy and Reimer (1987), who studied the NMR of phosphorus-31 dopant atoms in as-deposited films, then after annealing, and again after exposure to light. In as-deposited films nearly all the P atoms are threefold coordinated; after annealing, a significant fraction of the P atoms acquire fourfold coordination, but light transforms these back to threefold sites. Thus they found a link between metastability and dopants, one that was also found in studies of the doping dependence of both the built-in and light-induced densities of defects as measured by PDS (shown in Fig. 4.6) (Skumanich, Amer, and Jackson 1985), but that has not received much attention.

Transient Photocapacitance and Photocurrent

Transient photocapacitance and photocurrent measurements (Crandall 1981; Cohen et al. 1988; Cohen 1989; Nebel et al. 1989), have been shown to be useful for the study of defect states, being capable of providing information not otherwise readily available.

A detailed comparison has been made using transient photocapacitance and transient junction photocurrent measurements (Cohen et al. 1988) with the results shown in Figure 4.15. The photocapacitance spectra are very similar to spectra obtained by CPM and PDS, but with several significant differences:

1. Photocapacitance detects the net charge change in the depletion region rather than the absorbed energy, so that hole and electron transitions from gap states give signals of opposite sign.
2. The spatial region being measured can be varied in the photocapacitance measurement by varying the bias voltage.
3. The transient photocapacitance technique is several orders of magnitude more sensitive than PDS.
4. Varying the measurement temperature allows one to observe the competition between optical and thermal excitation processes.

The curves of Figure 4.15 have been overlapped below 0.9 eV, and attention is focused on the values at 1.2 eV. For this photon energy, photoexcitation of

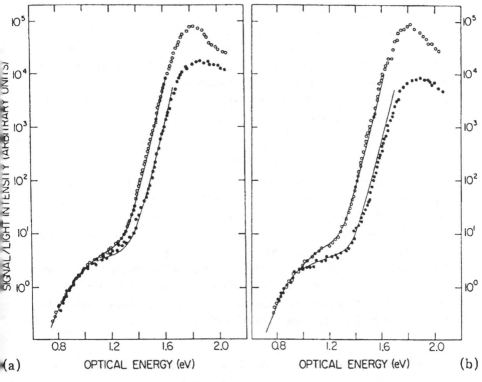

Figure 4.15 Comparison of photocapacitance spectra (●) and junction photocurrent transient spectra (○) at (a) 380 K and (b) 395 K. Superimposed solid lines are fits to the data of a model calculation. *Source:* Cohen et al. (1988).

an electron from a D⁻ center to the conduction band, and photoexcitation of an electron from the valence band to a neutral D^0 center to produce a D⁻ center and a hole, both contribute to the photocurrent; however, they give opposite changes in the depletion charge density, and hence in a photocapacitance measurement subtract from one another. Likewise, photoexcitation may excite an electron from a D^0 center to the conduction band, leaving a D^+ center; if the D^+ can thermally emit a hole on a time scale small compared to the measurement time, no change in photocapacitance will be observed, but a contribution to the photocurrent will be measured. In a similar way, photoexcitation from the valence band tail to the conduction band should give a positive contribution to the photocapacitance, but the remaining valence-band-tail hole may undergo rapid thermal emisson to the valence band, where it may leave the depletion region and produce no change in the photocapacitance, although the transient photocurrent signal will be enhanced. A comparison of the photocapacitance and photocurrent curves in the band-tail region of Figure 4.15

leads to the conclusion that 67 percent of the band-tail holes escape the depletion region at 380 K, whereas 84 percent escape at 395 K. From considerations such as this it was proposed that all the differences between the photocapacitance and photocurrent spectra in Figure 4.15 can be attributed to hole emission and/or transport processes. Application of these same methods to metastable defect creation indicates an intrinsic difference between light-induced and stable defects in a-Si:H.

Other Techniques

Other techniques that have proven useful include deep-level transient spectroscopy (DLTS) (Lang, Cohen, and Harbison 1982; Johnson 1983; Hack, Street, and Shur 1987), most useful in doped material; field effect measurements (Madan, LeComber, and Spear 1976; Jan, Knights, and Bube 1979, 1980; Goodman and Fritzsche 1980), which, however, may be strongly influenced by interface and surface effects; and space-charge-limited currents (Mackenzie, LeComber, and Spear 1982). The distribution of occupied defect states near the surface have been investigated using photoelectron spectroscopy in combination with a Kelvin probe (Winer, Hirabayashi, and Ley 1988).

4.6 Dependence of Fermi Energy on Defect Density

In undoped high-quality a-Si:H, the location of the Fermi energy at a given temperature is correlated with the value of the total defect density (Bube, Benatar, and Redfield 1994). If two of the three quantities – defect density, temperature, and Fermi energy – are known, the third is also known. This means that a change in the total defect density as a result of photoinducing or thermal annealing defects always produces a change in the Fermi energy in such undoped material, thus changing the occupancy of the defect centers and the details of phenomena dependent on this occupancy. No electronically related process can be interpreted accurately unless the effect of changes in defect density on Fermi energy are taken into account. Typical examples are the measurement of defect density by CPM (Bube et al. 1992) or by ESR that is sensitive only to the defects in the neutral state, or different variations with time of the measured defect density and the photoconductivity (Bube et al. 1994).

The correlation between Fermi energy and defect density was demonstrated by a large number of measurements of dark conductivity (and hence of the Fermi energy), over a range of defect densities and temperatures, obtained in the course of measuring the kinetics of optical degradation of undoped a-Si:H. The experimental situation was as follows:

1. All measurements of defect density were made at room temperature.

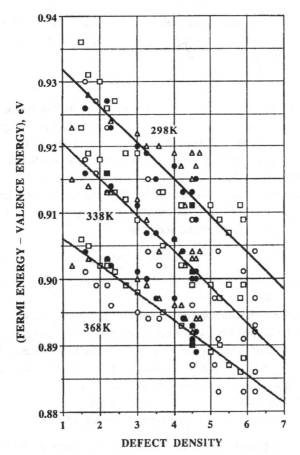

Figure 4.16 The Fermi energy as a function of the defect density (in units of 10^{16} cm^{-3}) at three different measurement temperatures: 298, 338, and 368 K. At each of these temperatures data are plotted from optical degradation runs at 45 °C (O), 80 °C (□), 95 °C (●), and 110 °C (△). The lines are empirical least-square linear fits to all the data at a fixed temperature. *Source:* Bube, Benatar, and Redfield (1994).

2. Defect density as a function of time was measured following optical degradation at each of four different temperatures.

3. After specific times of degradation, the dark conductivity was measured as a function of temperature. The Fermi energy was related (Bube, Benatar, and Redfield 1994) to the measured dark conductivity by $(E_C - E_F) = kT \ln(\sigma_0/\sigma_d)$, where E_C is the energy of the conduction-band edge, σ_d is the dark conductivity, and σ_0 was taken to be 150 S/cm after Stuke (1987; cf. Street 1991b, p. 232).

Figure 4.16 shows the variation of Fermi energy as a function of the defect density for measurements of conductivity at three different temperatures; at each of these temperatures data are plotted from each of the four optical degradation runs. At six different conductivity measurement temperatures, a relation is apparent such that

$$E_F = E_{F_0}(N) - \delta N \tag{4.4}$$

where N is the defect density at room temperature. At 298 K, for example, $E_{F_0} = 0.937$ eV and $\delta = 5.58 \times 10^{-19}$ eV cm^3. These values are independent of the history of the sample. A knowledge of the defect density to ± 10 percent enables one to specify the Fermi energy to within less than 0.5 percent (i.e., 4 meV out of 900 meV).

A similar relationship can be derived for the temperature dependence of the Fermi energy for a fixed value of the defect density, with results shown in Figure 4.17. Here a relation exists such that

$$E_F = E_{F_0}(T) - \varepsilon T \tag{4.5}$$

For a defect density of 2.2×10^{16} cm^{-3}, for example, $E_{F_0} = 1.026$ eV and $\varepsilon = 3.34 \times 10^{-4}$ eV K^{-1}.

The precision with which the defect density, the Fermi energy, and the temperature can be related makes this an interesting case for which to attempt to construct a model of the energy levels corresponding to the defects. This turns out to be one example among other indications that different kinds of DB defects, all with the same Si dangling-bond ESR signature, may be present in the material; other examples are given in Section 4.7.

A suitable model for the present data has a number of stringent constraints:

1. It must give quantitative agreement with the measured Fermi energy under some specified condition, e.g., $E_F = 0.91$ eV at 298 K for $N = 5 \times 10^{16}$ cm^{-3}.
2. It must give the observed dependence of Fermi energy on defect density at a fixed temperature (Fig. 4.16) and the observed dependence of Fermi energy on temperature at fixed defect density (Fig. 4.17).
3. Some account must be taken of the small temperature dependence of these energy levels because of the sensitivity of the data.

It is found that the Fermi energy decreases linearly with the decrease in band gap with increasing temperature, in such a way that the decrease in Fermi energy is about two-thirds of the decrease in band gap over the measured temperature range. In modeling it is assumed that the temperature dependence of the energy levels of the defects is the same as that of the Fermi energy.

The simplest model proposed to describe the variation in Fermi energy with defect density shown in Figure 4.16, while satisfying the conditions stipulated

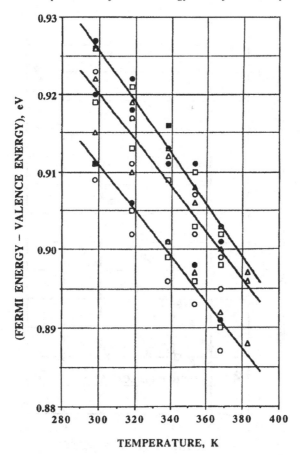

Figure 4.17 The Fermi energy as a function of the measurement temperature at three different values of the defect density (from top to bottom): 2.2×10^{16}, 3×10^{16}, and 4.5×10^{16} cm^{-3}. For each of these defect densities, data are plotted from optical degradation runs at 45 °C (O), 80 °C (□), 95 °C (●), and 110 °C (△). The lines are empirical least-squares linear fits to all the data for a fixed defect density. *Source:* Bube, Benatar, and Redfield (1994).

above, is one involving two different multivalent defects, each with discrete energy levels (measured as usual from the top of the valence edge) corresponding to $E_1(-/0) = 1.057$ eV and $E_1(0/+) = 0.857$ eV, and $E_2(-/0) = 0.965$ eV and $E_2(0/+) = 0.765$ eV, all values at room temperature. For simplicity it was assumed that $[E(-/0) - E(0/+)] = 0.2$ eV for both sets of defects. Comparison with the measured data yields the result shown in Figure 4.18, where values of the densities of the two sets N_1 and N_2 are plotted as a function of the total

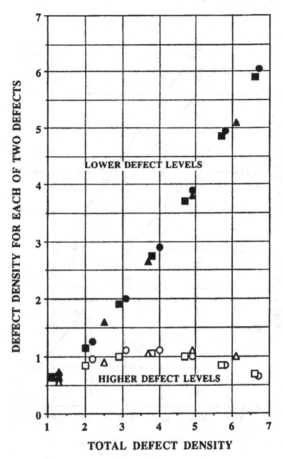

Figure 4.18 Calculated variation of the defect densities (all densities in units of $10^{16}\,cm^{-3}$) of two kinds of multivalent defects corresponding to the energy values given in the text. Values plotted are obtained for self-consistent solutions based on the lines drawn in Figure 4.16. Open points are calculated for the higher-lying defects, and solid points are calculated for the lower-lying defects. Measurements of Fermi energy were made at 298 K (O), 338 K (□), and 368 K (△). *Source:* Bube, Benatar, and Redfield (1994).

density ($N_1 + N_2$). Analysis of the data at each of three temperatures gives identical results. Both the higher-lying (N_1) and the lower-lying (N_2) defects show an increase in density at the beginning of optical degradation; but as optical degradation proceeds, the higher-lying defects reach a maximum and then decrease again, whereas the lower-lying defects increase continuously with degradation up to saturation.

4.7 Indications of Dangling-Bond Defects with Different Responses to Light and Heat

Dangling-bond defects with different kinds of response to light or heat might exist for a variety of reasons:

1. Defects may differ in the probability of formation or annealing by optical or thermal mechanisms, usually giving rise to the stretched-exponential form of the kinetics curves commonly observed.
2. Defects may also differ in the energy levels in the gap associated with their behavior, in their charge state, or in their capture cross sections.

Discussion of the first of these possibilities is given in detail in Section 5.5. In this section we give indications of the second.

There are also related issues to be considered:

1. Are dangling-bond defects present in thermal-equilibrium conditions in the material identical to DB defects formed by optical degradation?
2. Do different forms of excitation (e.g., light and electrons) produce DB defects with the same properties?

As indicated in the discussion in Section 5.5 of recent kinetics observations, analysis of electronic and optical properties in terms of only a total density of defects may be inadequate to deal with some situations. This is expected from the highly probable distribution of values of the properties of even a single set of defects, or from the possibility of more than one type of defect. Whatever may be the case, however, it appears that all the observed defects are associated with Si dangling bonds and have the same ESR signature with $g = 2.0055$.

There have been reports in which cells degraded at higher temperatures or for longer times are more difficult to anneal than those degraded at lower temperatures or for shorter times (Bennett et al. 1987). Similarly, in experiments in which the annealing at 150 °C of defects quenched in at 250 °C and 190 °C was compared, it was found that the additional defects introduced at the higher temperature annealed ten times faster (McMahon 1992).

Grimbergen et al. (1993) found that the saturated density of defects induced by electron-beam irradiation at room temperature was four to six times larger than the saturated density of photoinduced defects. The additional defects formed by electron-beam irradiation annealed rapidly at room temperature, unlike those induced by light.

Section 4.6 provides an example of evidence, based in that case on an analysis of the relationship between the defect density and the Fermi energy, for the existence of two different kinds of dangling-bond defects in a-Si:H with different, but closely spaced, energy levels. Other cases of this kind of evidence have been reported in the literature, and a few typical examples are summarized here. Each of the listed cases differs substantially from the others and cannot be directly compared; still, they suggest possible variations in defects induced by different methods.

Sakata et al. (1992) investigated the properties of defect states using analysis of CPM spectral-response curves, assuming that the energy states existed in Gaussian distributions of density with energy. They interpret their results to indicate the existence of two types of deep defect states in undoped a-Si:H. Tran et al. (1993) report results indicating that photoinduced metastable defects decrease the electron lifetime near 300 K more strongly than those defects existing in a thermal-equilibrium situation. Their results suggest that photo-induced defects have a larger hole-capture coefficient than defects produced in thermal equilibrium, and therefore serve as more effective recombination centers. Saleh et al. (1992, 1993) draw related conclusions from studies involving photoinduced ESR, CPM determination of defect density, and dark conductivity. They find that photoinduced defects differ from those formed during deposition or high-temperature annealing, the photoinduced variety corresponding to higher-energy states. Zhang et al. (1994) investigated the formation and annealing of defects induced by light at 77 K. Their results suggested a broad distribution of annealing activation energies, extending to lower energies at 77 K than at room temperature, and the defects with lower annealing activation energies appeared to be more effective recombination centers than those with higher annealing energies.

5

Hydrogenated Amorphous Silicon: Photoinduced Defect Kinetics and Processes

5.1 Kinetics of Generation and Annealing of Photoinduced Defects

In efforts to elucidate the nature and origin of the metastable defects induced by light in a-Si:H (Staebler and Wronski 1977), there have been many studies of the kinetics of their generation and annealing. Although such studies do not establish any microscopic models, they can provide significant guides as to which models may be acceptable. In the interpretation of such studies it is useful to be mindful of the various other ways of producing dangling-bond defects with similar properties. In particular, the observation that such defects can also be produced in the dark by passage of forward current in a $p–i–n$ device (Staebler et al. 1981) has led to the general belief that the defects are not produced by the light directly, but rather by the energy released when excess carriers recombine (Staebler et al. 1981) or are captured (Crandall 1991) at a localized center. Another significant observation is that light of photon energies down to and below 1 eV can cause this degradation even though the band gap is ≈1.75 eV.

The earliest observations of photoinduced degradation of lightly doped a-Si:H – the Staebler–Wronski effect – found that both the dark conductivity and photoconductivity decreased as a result of light exposure (see Fig. 1.2), and that these effects could be annealed away at about 150 °C for a few hours (Staebler and Wronski 1977). Thus there were two kinetic terms evident for the rate of change of the defect density N: a positive, defect-generating term that must contain the light intensity, and a negative, thermal annealing term. When the first two (nearly contemporaneous) quantitative measurements were made of the transient behavior during degradation (Lee et al. 1985; Stutzmann et al. 1985), these two kinds of terms formed the basis of the analyses of the kinetics; but even the simplest aspect of the kinetics, the description of ther-

113

mal annealing, was interpreted with different specific forms, one monomolec-
ular and one bimolecular. That conflict was later resolved with the recognition
that those curves have neither of these forms, but rather are stretched exponentials, which are characteristic of *dispersive* rate processes, that is, those having
a distribution of time constants instead of a single time constant (Kakalios,
Street, and Jackson 1987). A distribution of rates is expected in an amorphous
material because of the distribution in the details of the local structural environments of the defects, which should influence the quantitative values of local parameters, even for a single type of defect. Its consequences are described
below.

The first proposal of a particular microscopic model of defect generation
based on quantitative kinetics was that by Stutzmann, Jackson, and Tsai
(1985). In that formulation, defect generation was hypothesized to be caused
by the unmediated recombination (simple annihilation) of a free electron and a
free hole, so the generation term in the rate equation for N was taken to be
proportional to the product of the densities np of the free carriers. This model
led to the following rate equation for defect generation:

$$dN/dt = \varepsilon A \, np \tag{5.1}$$

where t is the duration of light exposure, A is the transition probability for
the np annihilation, $A \, np$ is the number of such transitions per unit time and
volume, and ε is an efficiency coefficient for creating dangling bonds. At room
temperature (at which most degradation studies are done), annealing was taken
to be negligible, so this was the only term used.

Because of the low photon energies capable of forming defects, it was assumed (Stutzmann et al. 1985) that the annihilating carriers are in the energy-
band tails of states near the bands. The carrier densities were each taken to be
proportional to the light intensity and inversely proportional to the existing
defect density N, which was assumed to determine the carrier lifetime by the
usual recombination-center processes. Thus, $n = G/A_n N$ and $p = G/A_p N$, where
G is the generation rate of photoexcited carriers (proportional to the light intensity), and A_n and A_p are the relevant transition probabilities. In this picture,
although the carrier lifetimes are dominated by the defects serving as recombination centers, it is not these recombination processes that produce more defects; only some small fraction of the total recombination events that occurs
by simple np annihilation can create defects. When these relations are combined and Eq. (5.1) is integrated, the resulting time dependence of the defect
density becomes

$$N(t) \approx c_{SW} \, G^{2/3} \, t^{1/3} \tag{5.2}$$

where the Staebler–Wronski coefficent c_{SW} contains several of the earlier constants, and the initial value of N is regarded as negligible. This result was found to fit degradation data over limited ranges (Stutzmann et al. 1985).

This model assumes that the defects are formed by breaking Si–Si bonds, and so has no provision for any maximum number of defects that could form. Also, with no defect-recovery process other than thermal annealing, according to this model there would be no end to this rise at low temperatures. This electron–hole formulation was widely accepted, although several problems have been pointed out (Redfield 1986a, 1989). Among these were the fact that extensive other degradation data (Lee et al. 1985) did not have this $(\text{time})^{1/3}$ dependence, and they did saturate at long times, even at 77 K, which is far below any thermal annealing temperature. Virtually no light-induced degradation measurements on good undoped a-Si:H have led to densities that exceed $\approx 3 \times 10^{17}$ cm^{-3}, implying that saturation at this low density occurs generally.

It was also pointed out that photoinduced defect reactions should be reversible, so that light must be able to cause recovery of defects as well as their formation (Redfield 1986b). This requires the presence of another term in the rate equation, one that also depends on the light intensity but is negative. This has major consequences because it leads to the occurrence of a steady-state condition in the presence of light even without any thermal annealing, and hence a saturation in the density of photoinduced defects even at low temperatures. This principle of reversibility was then applied to thermal processes as well, so that the rate equation for the defect density N was fully symmetrical and had four terms (Redfield 1988). It also meant that the generation and recovery aspects of defect behavior could not be separated as had been done previously. A maximum value of the density of possible defect centers N_T was also introduced in principle (Redfield 1988), although its value is not known. Thus an alternative rate equation was proposed to be

$$dN/dt = C_1 R(N_T - N) - C_2 RN + v_1(N_T - N) - v_2 N \qquad (5.3)$$

Here R is the total carrier capture (or recombination) rate, C_1 and C_2 are effectiveness coefficients for photoinduced generation and recovery of the metastable condition, and v_1 and v_2 are analogous coefficients for thermally induced generation and recovery, respectively. The quantity $(N_T - N)$ is the density of centers that are in their ground state and are thus available for defect formation. The dimensions of C_1 and C_2 are volumes, and v_1 and v_2 are frequencies; these latter are taken to be thermally activated with barrier energies E_1 and E_2 (shown in Fig. 4.4). One other difference of this formulation from that of Stutzmann et al. (1985) was that since R is the *total* rate, this implies that only one type of recombination process is significant: that occurring at the defects. It also follows that the electronic excitation rate G is proportional to R, from which some fraction produces (or restores) defects. It does not matter if the carrier lifetime changes during degradation because the carrier density will change proportionally, leaving R unaffected. This approach (Redfield 1988) is not based on any particular microscopic model of the defects or their transition processes; only the metastability of the centers is invoked, in the form of Figure 4.4. This has both advantages and disadvantages, since it need not assume a particular microscopic model, but neither can it confirm any model.

When $dN/dt = 0$, Eq. (5.3) has as its steady-state solution (or saturation) a value N_{sat} that is proportional to N_T and depends on the excitation rate $R(=G)$ and all four of the coefficients; N_{sat} depends on temperature through the thermally activated v_1 and v_2. The resulting properties of N_{sat} as a function of temperature and light intensity are shown in Figure 5.1 for two values of R and with a value of ΔE (defined in Fig. 4.4) of 0.14 eV taken from the work of Street (1987). Note that N_{sat} becomes independent of intensity and temperature at low temperatures, then drops and goes through a minimum at intermediate temperatures, and then rises at high temperatures as thermal effects become dominant and the effects of light become negligible (Redfield 1988). These steady-state results are fully consistent with the published data of Lee et al. (1985), and a number of other features of this behavior have been qualitatively confirmed in subsequent published measurements; however, in the light of recent reports of light-enhanced annealing, it appears that C_1 and C_2 are not simple constants as had been assumed (Redfield and Bube 1994).

The transient behavior was shown (Redfield 1988) to have the important property that an increase in light intensity shifts the response to shorter times, rather than raising the density at a fixed time. This difference becomes significant when evaluating an activation energy for defect generation by using various intensities and temperatures. A dependence of the time constant of response on temperature and light intensity was also predicted (Redfield 1988); this has been confirmed in part by subsequent observations.

Dispersive Behavior

The transient behavior $N(t)$ predicted by Eq. (5.3) is a simple exponential, however, which is known to be incorrect; it was only after dispersive attributes were introduced into the symmetric rate equation that good descriptions of both the steady-state and transient properties were obtained (Redfield and Bube 1989; Bube and Redfield 1989a). As mentioned above, this dispersive character and the resulting stretched exponential (SE) transients belong to disordered systems in general, since they naturally have a range of time constants

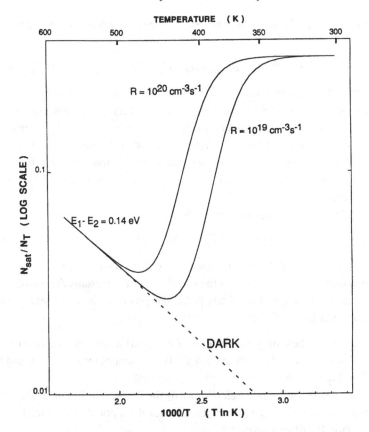

Figure 5.1 The steady-state defect density N_{sat} as a fraction of the maximum value of the density of possible defect centers N_T in an Arrhenius plot using two values of light intensity near the solar intensity. The dark behavior is the equilibrium defect density, which may not be achieved at lower temperatures because the relaxation rates become very long. *Source:* Redfield (1988).

in their rate processes. The basic properties of SEs and transitions in a distribution of independent defects were presented in Section 2.5 and are extended here in a form appropriate to a-Si:H. It had already been found that thermal relaxation of a-Si:H from a quenched-in nonequilibrium state follows an SE (Kakalios et al. 1987), and it was proposed (Redfield 1989) that defect generation should also have a dispersive character, particularly since the center of a log–log plot (such as was normally used for the time dependence of defect densities) of an SE with normal parameter values may approximate the (time)$^{1/3}$ dependence that Stutzmann, Jackson, and Tsai (1985) used to describe portions of their data. The rate equation given by Eq. (5.3) was extended (Redfield and Bube 1989; Bube and Redfield 1989a) in the simplest manner by assuming that all of the terms would have the same dispersive properties, leading to

$$dN/dt = (t/P)^{(\beta-1)} [C_1 R(N_T - N) - C_2 RN + v_1(N_T - N) - v_2 N] \quad (5.4)$$

where β is the *stretch parameter* related to the breadth of the distribution of contributing processes ($\beta < 1$), and P is a scaling factor that preserves the dimensions. The four coefficients no longer represent the physical properties of any particular center, but instead represent the distributions of the four processes. We note that although the derivation of the nondispersive form Eq. (5.3) is all first-principles kinetics, the dispersive form in Eq. (5.4) is phenomenological.

The solution of this equation is the SE

$$N(t) = N_{sat} - (N_{sat} - N_0) \exp[-(t/\tau)^{\beta}] \quad (5.5)$$

where N_0 is the initial density of metastable defects, and τ is an effective time constant for the transient; as with the coefficients, τ represents the distribution rather than the time constant of any particular process. There are several general properties of this form:

1. As β approaches unity the breadth of the distribution shrinks to zero, the dispersive character disappears, Eq. (5.4) degenerates into Eq. (5.3), and the SE in Eq. (5.5) becomes a simple exponential.
2. In the steady state (in which $dN/dt = 0$) the dispersive factor in Eq. (5.4) drops out, so this extension of the kinetics does not alter the steady-state behavior described above.
3. Eq. (5.5) describes both degradation and annealing, depending only on whether $N_{sat} > N_0$ or not.

Properties of Stretched Exponentials

In addition to the defect kinetics of DX centers in AlGaAs and a-Si:H, stretched exponential transients are seen in dielectric relaxation in polymers, mechanical relaxation in glassy materials, decays of trapped hydrogen in crystals, and so on (Plonka 1986; Rekhson 1986; Scherer 1986). Such descriptions are appearing with growing frequency, so we include a brief discussion of some of the properties of SEs. It must be emphasized that, since dispersive systems imply a distribution in the values of physical parameters of relevant processes, the use of a single numerical parameter like the total defect density to describe the system is necessarily incomplete without a description of the distribution as well. (An example of this need is shown later in Figure 5.18, in which the history of two samples with equal degradations is seen to affect the properties.) Nevertheless, the total density of defects is generally all that

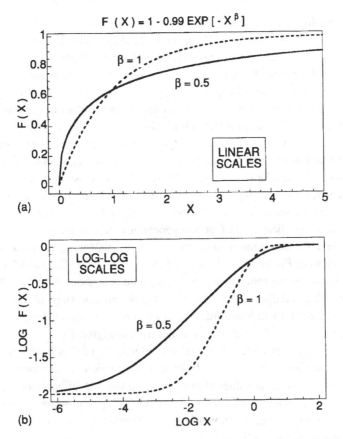

Figure 5.2. Comparsion of a simple exponential, $\beta = 1$, and a stretched exponential with $\beta = 0.5$. In the lower panel the same functions are plotted on logarithmic axes to illustrate the marked change in appearance caused by this commonly used choice.

is known, and there appear to be many cases in which it can serve as a useful guide; thus the SEs, which describe the behavior of this total, are of interest. One goal of further research is to learn also about the distributions.

A comparison between the simple exponential and SE is instructive, and is presented graphically in Figure 5.2 for these functions with the same initial and final values for both curves and a final/initial ratio of 100; the abscissa $X = t/\tau$ is a normalized time. Here the value of β distinguishes the two, since $\beta = 1$ corresponds to the simple exponential; the value $\beta = 0.5$ for the SE is chosen here for comparison because it is observed often in a-Si:H transients. It is clear that for all values of β the two curves are equal at the value $X = 1$, and for $\beta < 1$ the SE changes more rapidly at shorter times and more slowly at longer times; the word "stretched" obviously applies to the longer times only.

In fact, the slope of SEs $\to \infty$ as $X \to 0$ for all values of $\beta < 1$; this is a distinctive feature that has appeared in several sets of experimental data.

A logarithmic time scale is frequently used to permit display of a wide range of times, and for a-Si:H a logarithmic ordinate is often used as well. For that reason, the bottom panel of Figure 5.2 is presented to exhibit the striking differences in appearance that the change in scales create. Most of these are due to the change in the use of a logarithmic time scale, which stretches out the shorter times. Examples of these two types of presentations of experimental data appeared in Figures 4.13 and 4.14, and it was not apparent at first that these two are so similar or that they represent SEs. One aspect of such curves significant for the interpretation of experimental results is that the use of a logarithmic time scale creates an inflection point in the curve that has no special physical significance. In fact, the appearance of a nearly straight portion around the inflection point of such curves can be misleading, as it was in the interpretation of Figure 4.14. It was later shown (Redfield 1989) that for such curves, in which the ratio of final to initial values is about 20 (as it often is for light-induced defect densities in a-Si:H), the apparent slope of this nearly straight section is about one-third; this was given an unwarranted physical significance as a result of the choice of scales used in Figure 4.14.

Despite their prevalence, physical interpretations of SEs have generally eluded researchers. Their central feature, a distribution of time constants τ, arises from nonidentical values of physical parameters at different locations in a material. An expression of the way in which they contribute to observed effects is the following, which would apply to the total density of defects $N(t)$ as used earlier in this section:

$$N(t) \approx \int_0^\infty f(\tau) \exp\left(\frac{-t}{\tau}\right) d\tau \qquad (5.6)$$

Here the function $f(\tau)$ is a spectral density (or distribution function) of the contributing time constants, and it is assumed that some physically necessary cutoffs in the range of τ are unimportant. Beyond this expression there is no general further interpretation, largely because the characteristics of different materials are quite different from each other. In many cases the materials (e.g., glasses) have processes that are strongly coupled throughout the material, and individual processes cannot be distinguished. In dilute, localized defects in semiconductors, however, there is little interaction among different contributing members of the set of metastable defects; they may be treated as independent (Redfield 1992). Then the interpretation simplifies greatly, making N a sum of subsets of defects, each subset having its own value of τ (Redfield 1992). This is now the interpretation of observed transients in DX properties in AlGaAs described in Chapter 2.

In these cases it is useful to exploit the fact that the right-hand side of Eq. (5.6) is exactly the Laplace transform of the distribution function $f(\tau)$; by calculating the inverse Laplace transform of the observed $N(t)$ we can deduce the function $f(\tau)$ (Plonka 1986). This inverse can be calculated in closed form in special cases, one of which is when $N(t)$ is an SE with $\beta = \frac{1}{2}$, for which

$$f(\tau) = \tfrac{1}{2}(\pi\,\tau\,\tau_0)^{-1/2}\exp(-\tau/4\tau_0) \qquad (5.7)$$

where τ_0 is the effective time constant of an observed SE (Rekhson 1986). This was applied to annealing relaxation curves of a-Si:H, for which this value of β has been found to be appropriate (Redfield 1992). In such cases the annealing time constants are taken to have thermally activated forms whose controlling physical parameter is the activation energy over a barrier E_{ac}. Then the distribution of these barrier heights can be inferred from the distribution of time constants by the general relation for activated processes:

$$g(E_{ac}) = f(\tau)\,(d\tau/dE_{ac}) = f(\tau)\,\tau/kT \qquad (5.8)$$

The results of this procedure are shown in Figure 5.3, whose abscissa represents either the time for the SE or the time constant in Eq. (5.7). The solid curve then shows the distribution both of the time constants on the bottom scale and of the derived activation energies in Eq. (5.8) along the top. Thus we find an unusual case in which experimental data can be used to obtain information on the distribution of values of a physical parameter.

Use of Stretched Exponentials

This SE formulation of defect generation has been broadly successful in describing experimental observations from many laboratories. One of the most significant sets of data against which to test models of degradation kinetics was provided by measurements of Park, Liu, and Wagner (1989). By using either a high-intensity krypton laser or ordinary light on two different samples, they were able to demonstrate clearly that there is indeed a saturation of the defect density, and that its value is independent of light intensity, as had been predicted (Redfield 1988). Transient data (Park et al. 1989) were later shown to be fitted well by SEs like Eq. (5.5), and changes in intensity caused changes in the effective time constants rather than in the defect densities (Bube, Echeverria, and Redfield 1990). A different type of SE description for degradation kinetics, but one based on the Stutzmann, Jackson, and Tsai model, was developed by Jackson (1989); this formulation, however, disagrees with the data of Park, Liu, and Wagner (1989) by predicting a variation of the saturated defect density with light intensity, and no variation in the time constant with intensity.

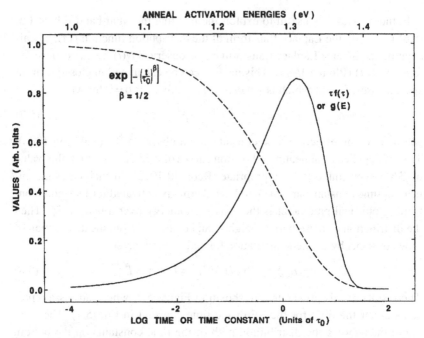

Figure 5.3 Dashed curve: a stretched-exponential relaxation with $\beta = 0.5$. Solid curve: the inferred distribution of time constants (when the bottom scale is used) or of activation energies (when the top scale is used and the temperature is 400 K). *Source:* Redfield (1992).

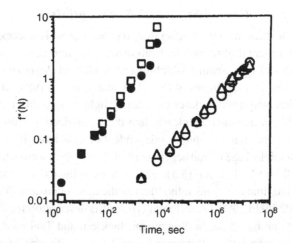

Figure 5.4 The data of Park, Liu, and Wagner (1989) on two samples at two light intensities using the form of Eq. (5.9) to test the closeness of the data to true stretched exponentials, which produce straight lines. The left-hand pair are for high-intensity light. *Source:* Bube, Echeverria, and Redfield (1990).

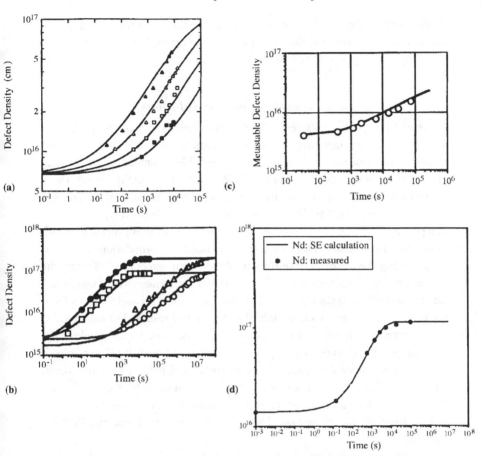

Figure 5.5 Fits of various experimental degradation data with stretched exponentials. *Sources:* Figure from Redfield and Bube (1991b); data from (a) Stutzmann, Jackson, and Tsai (1985), (b) Park, Liu, and Wagner (1989), (c) Curtins et al. (1988), and (d) L. E. Benatar (1990 unpublished data).

Also shown (Bube et al. 1990) was a convenient method for determining the precision of an SE fit by rewriting the exponential part of Eq. (5.5) as $\exp[-Kt^\beta]$, so that

$$\log f^*(N) = \log K + \beta \log t \qquad (5.9)$$

where $f^*(N) \equiv \ln[(N_{sat} - N_0)/(N_{sat} - N)]$. For a true SE, a plot of $\log f^*(N)$ vs. $\log t$ yields a straight line with slope β and intercept $\log K$. This method was applied to the data of Park et al. (1989) with the results shown in Figure 5.4.

The SE description of light-induced degradation was tested further by applying it to several published sets of degradation data (including those of Stutzmann et al. [1985] and Park et al. [1989]) with results collected in Figure 5.5

(Redfield and Bube 1991b). For each sample one parameter was adjusted once because of uncertainties in the values of the coefficients; the variations caused by intensity changes followed from the equations.

A Comparison of Formulations

In view of the marked differences in the formulations of Eqs. (5.2) and (5.5) and their resulting descriptions, an investigation was made of a variety of mathematical models representing various possible combinations of physical processes for optical degradation in a-Si:H (Bube and Redfield 1989a). These all were taken to have carrier lifetimes inversely proportional to the defect density. Nine models were analyzed, identified by (1) the assumed type of process that produces metastable defects and (2) the assumed type of process that controls the carrier lifetime. Dispersive and nondispersive formulations were tested, including the nondispersive form used by Stutzmann et al. (1985) and the dispersive one of Redfield and Bube (1989). For the former case, it was necessary to add both a maximum possible density of centers N_T and a light-induced recovery term to permit simulation of experimental results. For each model, calculations were made of the predicted low-temperature kinetics and the dependence of the saturation density on temperature and intensity. The results of this investigation were that published data could be simulated with the extended version of the Stutzmann et al. (1985) formulation only if dispersive effects were neglected; if dispersion were included, however, only the formulation of Redfield and Bube was able to fit reported data (Bube and Redfield 1989a).

A detailed examination of subsequently published kinetics data has led to the conclusion that there is no fundamental physical mechanism to be directly associated with a (time)$^{1/3}$ behavior (Bube et al. 1990). It had been shown (Redfield 1989) that a log–log plot of an SE has an inflection point near the middle of its range that produces a nearly straight portion, which can easily be mistaken for a power-law behavior, and that in common cases the "power" appears to be about one-third. It was also shown (Bube et al. 1990) that the slope of a real SE simply goes through a peak in this region, and that for the data of Park et al. (1989) the peak values may be near one-third, but actually vary with temperature.

These conclusions were confirmed later with a different method of comparison of the data of Park et al. (1989) with the (time)$^{1/3}$ hypothesis and the SE description (Bube and Redfield 1992). There the role of temperature changes in both the $t^{1/3}$ and SE models was analyzed, with the conclusion that the ability of both these models to describe the data at room temperature is a coincidence;

at any other temperature, the $t^{1/3}$ model is unable to describe the data. Specifically, whenever analysis of the kinetic data for light-induced formation calls for a β that is temperature-dependent, as is commonly the case, the use of a "power-law" model is found to require that the power vary with temperature.

5.2 Pulsed-Light-Induced Defects

In investigating the formation of metastable dangling-bond defects in a-Si:H from various conditions of preparation, doping, and processing, it is desirable to have available accelerated degradation techniques so that the saturation density of DB defects can be determined in a reasonably short time. The most common of these is simply the use of a high-intensity light source, for example, from a Kr laser (Park et al. 1989). Several other techniques have also been tested for this purpose, including pulsed-light excitation (Rossi, Brandt, and Stutzmann 1992), kilo-electon-volt electron irradiation (Schneider and Schroeder 1989a), and current injection (Street and Hack 1991). Sections 5.2–5.4 focus on these additional methods for rapid degradation.

Some general considerations apply for any method of high-intensity excitation. As the power of an excitation beam increases, sample heating and control of the sample temperature often become a problem that precludes the independent measurement of thermal and intensity effects. If the response of the defect generation is linear in intensity, then increasing the intensity I to accelerate degradation cannot avoid this problem. However, it has been shown in the discussion of the kinetics of defect generation that, over the ranges studied optically, the effective time constant varies as $\tau_{\text{eff}} \approx I^{-(1/\beta)}$, where β is the stretch parameter of the SE describing the generation kinetics (Redfield and Bube 1989). Since β is generally considerably less than 1 (often about 0.5), the effective time constant decreases rapidly as I increases. As long as this applies, there is a benefit from the use of high-intensity excitation. By the use of light pulses with a low duty cycle, heating can be minimized while accelerating the degradation.

Investigations of the degradation kinetics have been carried out using pulses from a flash lamp (Nevin, Yamagishi, and Tawada 1989) as well as from lasers (Stutzmann et al. 1991a,b). The first observation of saturation of the defect density and of the photoconductivity in a-Si:H using laser pulses (640 nm) was reported by Hata et al. (1992). Using identical samples they compared the effects of degradation to saturation by continuous-wave (cw) light from a xenon arc lamp or by light pulses from a dye laser. Degradation by both techniques resulted in saturation at a metastable defect density of about 10^{17} cm^{-3}.

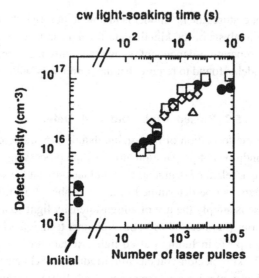

Figure 5.6 Defect density as functions of the number of 5 mJ cm^{-2} nanosecond light pulses (bottom abscissa) at the optical degradation temperatures of 25 °C (□) and 100 °C (●), and as a function of 3 W cm^{-2} cw optical degradation (top abscissa) at 35 °C (◇). Also shown (△) is the saturation level under 3 W cm^{-2} cw light soaking at 110 °C. *Source:* Hata et al. (1992).

Typical results are shown in Figure 5.6, where the defect density determined by CPM is plotted both as a function of the number of light pulses for the pulsed photoexcitation and as a function of the photoexcitation time for the 3-W cm^{-2} cw photoexcitation, for degradation at the indicated temperatures. A somewhat smaller saturation value is indicated for cw photoexcitation at 110 °C than at 35 °C, consistent with the observation that the saturation value at higher temperatures does depend on the actual photoexcitation rate, and this rate for pulsed photoexcitation is greater than that for cw. It is possible to calculate an "effective photoexcitation time" suitable for comparing the two modes of photoexcitation (e.g., Fig. 5.6 indicates that the "effective photoexcitation time" is the same for 100 s of cw-light excitation and ten laser pulses for the intensities used) (Hata et al. 1992).

Hata et al. (1992) also compare data on the optical degradation of photoconductivity using nanosecond-pulse photoexcitation, with the data obtained from femtosecond-pulse and cw photoexcitation (Stutzmann et al. 1991a), as a function of the "effective photoexcitation time." All of the data fall on essentially the same curve. The results indicate that the failure to observe saturation of the photoconductivity under these conditions (Stutzmann et al. 1991a) was the re-

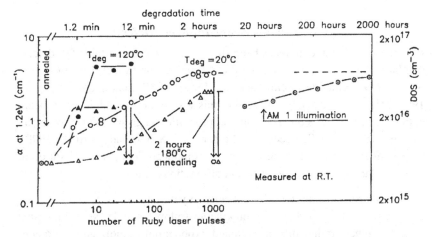

Figure 5.7 Dependence of the density of states (DOS) and subgap α (1.2 eV) on time, or number of ruby laser pulses. Sample A, circles; sample B, triangles; room temperature degradation, open points; degradation at 120 °C, filled points. AM1 degradation for sample A is shown for comparison. *Source:* Kocka, Stika, and Klima (1993).

sult of terminating the experiment at values of the "effective photoexcitation time" too short to see the saturation.

In order to obtain a light source for pulsed photoexcitation experiments that would be useful in giving uniform degradation across the thickness of the sample with thicker a-Si:H films, Kocka, Stika, and Klima (1993) investigated the use of a pulsed ruby laser (694 nm). Two undoped samples ($\approx$2 μm thick) were investigated, sample B having been prepared, however, under a preprocess vacuum at least one order of magnitude better than sample A. The overall results shown in Figure 5.7 indicate rather different behavior for samples A and B. In particular, at room temperature, the density of states in sample A is three times larger after ten pulses than in the annealed state, but almost no change is observed for sample B; also, the saturated density for sample B is appreciably lower than that for sample A. Similar differences are observed between the two samples for optical degradation of the dark conductivity. Kocka et al. (1993) propose an explanation involving the assumption that sample A contains an appreciable concentration of O or N impurities (but $<10^{20}$ cm^{-3}, since no oxygen-related peak is observed in the infrared) forming energy levels within the valence-band tail, and that photoactivation of these impurity-related states leads to an increase in E_0, the slope of the valence-band tail, and a sharp downward shift of the Fermi energy.

5.3 Electron-Beam-Induced Defects

Electron-beam degradation has been shown to produce metastable DB defects similar to those caused by strong optical illumination (Street, Biegelsen, and Stuke 1979; Schneider, Schroeder, and Finger 1987; Scholz and Schroeder 1990). The defect density produced in this way saturates with electron-beam exposure and depends on such material properties as growth temperature and band gap (Schneider, Schroeder, and Finger 1987; Scholz and Schroeder 1990), very similar to the effects shown by photoinduced degradation (Park et al. 1990a). Major differences described in this section are that

1. the electron beam produces a higher density of defects,
2. electron-beam defect-generation kinetics follow a simple exponential time dependence, rather than a stretched exponential (Scholz and Schroeder 1991), and
3. the higher density of defects induced by an electron beam undergoes faster relaxation than the defects formed by light.

Since there is such a large energy difference between visible photons and 20-keV electrons, differences between the degradation processes might well be expected. The highest generation rate under cw optical illumination is $\approx 10^{22}$ $cm^{-3} s^{-1}$, whereas for an electron flux density of 10^{19} $cm^{-2} s^{-1}$ absorbed uniformly in 2-μm thickness, the generation rate is $\approx 10^{26}$ $cm^{-3} s^{-1}$, about 10^4 times larger, but with little heating.

Background

A summary of the some of the earlier experiments involving electron-beam irradiation of a-Si:H is given by Schade (1984), who describes particularly effects on photoluminescence and infrared absorption. The threshold for electron-beam-induced damage in a-Si:H is only 1 keV, much lower than the 100 keV required for crystalline silicon. Thus defect formation in a-Si:H cannot be the result of normal atom displacements. The effective absorption associated with electron irradiation can be varied from essentially homogeneous if megavolt electrons are used (Street, Biegelsen, and Stuke 1979; Voget-Grote et al. 1980), to essentially the same as for photons if kilovolt electrons are used (Voget-Grote et al. 1980).

Defect Formation in Thin Films of Undoped a-Si:H

Scholz and Schroeder (1991) examined the effect of electron intensity on degradation kinetics and the saturation level by irradiating different parts of the

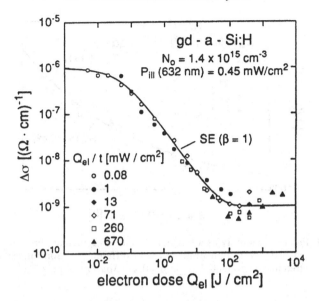

Figure 5.8 Photoconductivity as a function of the electron dose of 20-keV electrons for a glow-discharge-deposited a-Si:H film irradiated with different electron intensities Q_{el}/t. Curve is a simple exponential (i.e., $\beta = 1$ if written in stretched-exponential form). *Source:* Scholz and Schroeder (1991).

same sample with different 20-keV electron intensities Q_{el}/t, where Q_{el} is the electron dose in joules per square centimeter and t is the time, so that the units of Q_{el}/t are given in milliwatts per square centimeter. Figure 5.8, which shows the dependence of photoconductivity $\Delta\sigma$ as a function of Q_{el} deposited on the sample using six different electron intensities, illustrates the following:

1. The defect creation is independent of the irradiation intensity over four orders of magnitude.
2. The state of degradation is determined only by the total number of electrons incident on the sample.
3. For electron doses higher than 70 J/cm^2 saturation is observed.
4. If a stretched-exponential description is applied, the best agreement is achieved with $\beta = 1$, indicating a simple-exponential kinetics.

These properties indicate a linear response to intensity over this range, in contrast to the nonlinear dependence on light intensity.

Figure 5.9 shows the dependence of the photoconductivity as a function of annealing time at 296 K after six different irradiations (Scholz and Schroeder 1991). It also shows that the defect annealing is independent of the irradiation

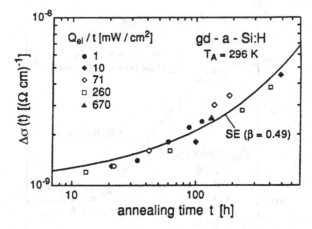

Figure 5.9 Photoconductivity as a function of annealing time at 296 K for a glow-discharge-deposited a-Si:H film irradiated with different electron intensities Q_{el}/t. Curve is a stretched exponential with $\beta = 0.49$. *Source:* Scholz and Schroeder (1991).

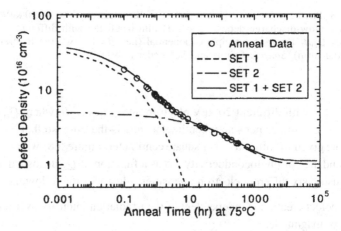

Figure 5.10 Relaxation of electron-beam-induced defect density at 75 °C, fit to the sum of two stretched-exponential relaxation curves. *Source:* Grimbergen et al. (1993).

intensity, but now an SE with $\beta = 0.49$ represents the best fit to the data, almost the same value as reported for the thermal annealing of light-induced defects (Kakalios et al. 1987). The results were interpreted as being consistent with the earlier assumption that kilo-electron-volt electron irradiation produces defects by a process that leads directly to the dissociation of a Si–H or a weak Si–Si bond (Gangopadhyay, Schroeder, and Geiger 1987). The saturation density of defects was found to be independent of electron intensity over three orders of magnitude (Scholz and Schroeder 1991).

A quantitative comparison has been carried out by Grimbergen et al. (1992) of the defect densities and characteristics produced by 20-keV electron beam irradiation with those produced by strong 1.9-eV light on similar undoped a-Si:H samples. The saturation defect density produced by electron irradiation at 300 K is within 5 percent of the value obtained at 225 K, but is about four to six times larger than that produced by photoexcitation. The independence of the saturation density of temperature in the low-temperature range shows it to be the result of a limited number of possible defects, and not simply a steady state between electron-beam creation and thermal annealing of defects.

Both electron irradiation and photoexcitation cause a considerable decrease in photoconductivity, but qualitatively different results are indicated for the dark conductivity, which decreases for photoexcitation but actually increases for electron-beam irradiation (Voget-Grote et al. 1980).

Relaxation of Defects in Thin Films of Undoped a-Si:H

After defects have been formed by electron irradiation, the relaxation of the defect density with time at different temperatures has been observed (Grimbergen et al. 1993). Relaxation curves for annealing at 22 °C, 50 °C, and 75 °C were each approximately fit by a stretched-exponential curve. The SE relaxation time constant is strongly affected by annealing temperature and shows an activation energy of 0.9 ± 0.1 eV, essentially the same as for relaxation of light-induced defects (Jackson and Kakalios 1988), although the values of the relaxation time constant are much smaller, decreasing from 50 s at 22 °C to 0.2 s at 75 °C.

The detailed shape of the measured relaxation curves for 75 °C, however, and the high final density of the fitted stretched exponentials (especially at 22 °C) indicate that relaxation is incomplete and is not fully described by a single SE. The relaxation appears to follow at least two stages, corresponding to two different dominant sets of defects: a large initial density that relaxes rapidly (Type 1), followed by a slower relaxation with a time constant more characteristic of light-induced defects (Type 2). The data can be described by the sum of two SEs with different time constants, as shown in Figure 5.10 for data corresponding to annealing at 75 °C. Type 1 defects correspond to defects formed by the electron irradiation uniquely, whereas Type 2 defects correspond to those induced also by photoexcitation. Both Types 1 and 2, however, are apparently the same kind of dangling-bond defects, as identified by ESR (Dersch, Schweitzer, and Stuke 1983). There is some evidence that Type 1 defects may be associated with the a-Si:H/glass interface.

Comparison of the magnitude of the saturation density between photo-induced and electron-beam-induced irradiation can be meaningfully carried out provided that the additional density of defects with fast relaxation, associated with electron irradiation, are taken into account.

Quantitative interpretation of the relaxation of the photoconductivity – a larger decrease in photoconductivity being caused by electron irradiation than the increase in defect density as measured by the CPM method – is complicated by the need to take detailed account of the effect of the motion of the Fermi level during relaxation on the value of the measured CPM density, as is also the case for light-induced defects (see Sections 4.5 and 5.7) (Bube et al. 1992).

Defect Formation in Thin Films of Doped p- and n-Type a-Si:H

The degradation of *n*-type phosphorus-doped and *p*-type boron-doped films of a-Si:H has been investigated by Scholz, Schehr, and Schroeder (1993) using 20-keV electron irradiation. The saturation value of the defect density is approximately the same as that found for electron-induced defects in intrinsic a-Si:H, and the kinetics are described by a simple exponential, as is also the case for the intrinsic material. These results suggest that the direct formation of defects in the a-Si:H by electron irradiation is not influenced by the presence of the dopant atoms. It was concluded that the degradation of photoconductivity and the increase of defect density in doped samples due to electron irradiation are similar to those in intrinsic a-Si:H with equivalent parameters, regardless of the incorporated dopants.

The variation of dark conductivity in the initial and degraded states as a function of the location of the initial Fermi level ($E_{F,init} - E_V$), as determined by the dopants present, is shown in Figure 5.11. For initial Fermi-level positions near midgap, only small changes in dark conductivity occur with degradation (a small increase is indicated, as observed also above). For initial Fermi-level positions below midgap (*p*-type material) a relatively small decrease in dark conductivity is caused by electron irradiation, whereas for Fermi-level positions above midgap (*n*-type material) a much larger decrease in dark conductivity with irradiation is found.

Measurements on Amorphous Silicon Solar Cells

The degradation of a-Si:H *p–i–n* solar cells by 20-keV-electron irradiation has been investigated by Schneider and Schroeder (1989b) and by Scholz et al. (1993). Saturation of degradation was achieved in which the conversion efficiency of the cell was reduced to approximately 10 percent of its initial value, a much more severe effect than that caused by 1,000 h of AM1 illumination.

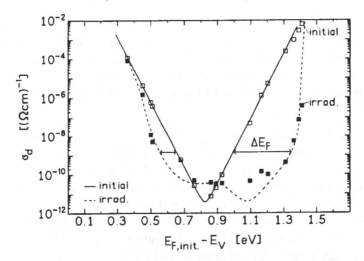

Figure 5.11 Dark conductivity of doped a-Si:H in initial and irradiated state as a function of the initial Fermi-level position. The dotted line is the result of a model calculation. *Source:*Reprinted from *Solid State Communications* 85, 753, Scholz, Schehr, and Schroeder, "Metastability in *p*- and *n*-Type a-Si:H Investigated by keV-Electron Irradiation," © 1993, with kind permission from Elsevier Science Ltd., The Boulevard, Langford Lane, Kidlington OX5 1GB, UK.

It was found that the relative reduction of the open-circuit voltage was generally larger than the relative reduction of the short-circuit current, but that the major effect of degradation was due to a reduction in the fill factor, caused either by an increase of bulk recombination or by interface recombination, or by both. Although the reduction in fill factor is similar for both electron irradiation and AM1-light exposure, the reduction in open-circuit voltage is much larger for the electron irradiation case. This difference appears to be related to a difference in the spatial distribution of the induced defects, associated with the energy dissipation of kilo-electron-volt electrons, in the *p–i–n* structure by the different methods of degradation.

5.4 Defects Induced in the Dark by Carriers

As mentioned earlier, excess densities (i.e., those above the minimum background number) of metastable defects can be induced by means other than light or electron beams. Two methods treated in this section are the use of current flow in *p–i–n* device structures (Staebler et al. 1981) and application of an electric field by a gate electrode in a field-effect-transistor (FET) structure (Hepburn et al. 1986). Such FETs of a-Si:H are coming into common use as the controlling thin-film transistors in active-matrix liquid-crystal displays and are discussed in Section 7.5. There is general agreement that the defects generated by both of these methods have the same basic character as those generated by light – that is, they are dangling-bond defects. Hence, a full understanding of metastable defects should encompass the phenomena in these observations too, and they may provide additional information on the defect properties. For example:

1. The ability of current flow in the dark to generate the same kind of defects as by light has led to the conclusion that the mechanism of light-induced defect formation does not involve direct interaction with light, but is driven rather by the energy released by excess carriers upon their recombination or capture at a localized center (Crandall 1987).

2. The FET observations of defect generation, however, seem to occur under conditions in which there is very small current flow, and hence small recombination rates (van Berkel and Powell 1987).

This latter result is consistent with data showing that the defects could be formed by the presence of either electrons or holes separately (Crandall 1987). Thus actual electron–hole recombination is not necessary; capture of a single charge or just the presence of excess carriers is sufficient to produce a defect.

Defects Induced by Current

Since the initial report (Staebler et al. 1981) that defects form in *p–i–n* structures of the type forming solar cells as a result of current flow in the dark, there have not been nearly as many kinetics studies using this method of degradation as by light, despite the fact that a wide range of excitation rates can be covered easily. One likely reason for this reluctance is that these structures can possess more complicated properties than the homogeneous films used with optical degradation. There has been, in fact, some controversy over the extent to which solar-cell degradation can be explained by the defects generated in the *i*-layer (see Section 7.4). Crandall (1987) performed a number of measurements on current-induced defects, but most were at high temperatures at which some differences in behavior are noted compared to the more common room-temperature data. Nevertheless, it has been found that the defects produced by light and current are similar, as indicated by ESR, annealing properties, and, to some degree, degradation rates. Implicit in such results is the conclusion that solar-cell degradation is dominated by the defects in the *i*-layer.

New results on the kinetics of recovery of current-induced defects were reported, however, by Street (1991a; cf. Street and Hack 1991), and are discussed further in Section 5.5. In Street's work it was found that after severe degrada-

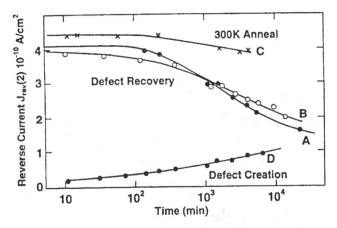

Figure 5.12 Recovery of current-induced defects following degradation at 1 A/cm^2 for about 1 h. Curve C is the recovery history at 300 K without current; curves A and B show the enhanced recovery caused by current flow (curve A at constant 1.7 V, and curve B at constant 5 mA/cm^{-2}). Curve D is the defect formation from an annealed state at a current of 5 mA/cm^{-2}. *Source:* Street (1991a).

tion by high forward current, the normally very slow recovery at room temperature is significantly accelerated by application of low currents during recovery, as shown in Figure 5.12. It was suggested (Street 1991a) that this observation may be related to the light-enhanced annealing of heavily degraded solar cells at elevated temperatures that had been reported by Delahoy and Tonon (1987). Such excitation-enhanced annealing of defects was soon observed with light-induced defects in homogeneous films, and the kinetics are discussed in Section 5.5. The central point of interest is that none of the existing rate equations for defect generation and annealing was capable of explaining such behavior, since their formulations all required *more* defects in the presence of excitation (Street 1991a).

Defects Induced in Accumulation Layers

When a voltage is applied to the gate electrode of an FET, charges become trapped near the a-Si:H/dielectric interface. This effect was studied (Hepburn et al. 1986) under conditions in which no current flowed between the source and drain. The gate voltage was removed, and after some variable delay time t_d a light pulse caused the remaining trapped charge Q to be cleared from the device. The amount of this remaining charge was measured as a function of the delay time and the sample temperature, thus providing data on the relaxation properties of the charge. From these results, the initial (unrelaxed) amount of

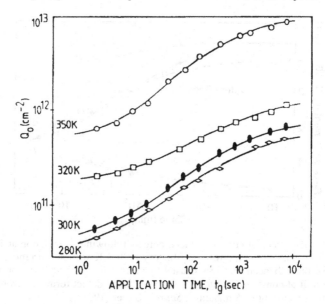

Figure 5.13 Influence of the duration t_g of the gate field on the magnitude of the trapped charge at zero delay time of the discharging light. *Source:* Hepburn et al. (1986).

charge Q_0 could be inferred, and that was studied as a function of the duration t_g of the inducing gate voltage, with the results shown in Figure 5.13 for four temperatures.

Interpretation of these results is complicated by several factors, the most important being that it is difficult to distinguish between effects in the a-Si:H and charge trapping in the dielectric – which had been invoked earlier to explain the first observed instabilities in such devices (Powell 1983). Nevertheless, analysis concluded that this field-induced effect forms metastable defects in the a-Si:H itself (Hepburn et al. 1986). This was supported by the agreement of the inferred density of states and annealing properties of these field-induced defects and those of light-induced defects of the familiar kind.

The possibility that observed field-induced charge-trapping effects in FETs could occur in either the dielectric or the a-Si:H caused continuing controversy in their interpretation, and an ingenious set of experiments was performed to isolate the two types of effects (van Berkel and Powell 1987). These experiments used *ambipolar* devices, which show either electron or hole accumulation layers depending on the polarity of the gate voltage, with some results shown in Figure 5.14. The observed asymmetry of the results of opposite polarities led to the interpretation that, although some charge is trapped in the

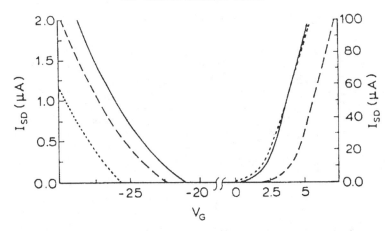

Figure 5.14 Transfer characteristics (source-drain current vs. gate voltage) for an FET. Solid lines, after anneal; dashed lines, after positive bias of 25 V for 4 h; dotted lines, after negative bias of –25 V for 4 h. *Source:* van Berkel and Powell (1987).

dielectric, there must be a contribution from defects in the a-Si:H also. This was the first convincing demonstration of such a conclusion, which was generally surprising because these defects form "without the need for a recombination event" (van Berkel and Powell 1987).

5.5 Recent Kinetics Results

As measurement techniques have advanced, several more complicated properties of the kinetics of defect generation and annealing in a-Si:H have been reported. Although they have been found in different ways, they have several common features that would be desirable to incorporate into analytic models of defect behavior. This is currently being attempted, but without consensus.

One of these newer properties appeared in Street's (1991a) current-induced-defect studies in p–i–n devices described in Section 5.4. The new observation there was accelerated relaxation of a high density of defects (that had been produced by strong excitation) caused by the presence of moderate levels of excitation. Such behavior could not be taken into account by any of the existing rate equations for defect kinetics described in Section 5.1.

The use of light to enhance annealing in homogeneous films was reported independently by Meaudre and Meaudre (1992a,b), who noted the similarity to Street's results, and by Isomura and Wagner (1992). Meaudre and Meaudre (1992a,b) formed defects by thermal quenching of their samples, and annealing rates were measured in the presence of various light intensities L and compared

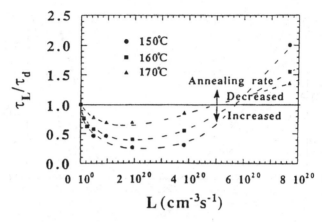

Figure 5.15 Relaxation time constants in the light τ_L compared to that in the dark τ_d at various intensities and at three temperatures. *Source:* Meaudre and Meaudre (1992b).

to the rate in the dark, with results shown in Figure 5.15. The relaxation times under light τ_L are shorter than in the dark τ_d for low intensities, but longer at the highest intensities.

In other detailed measurements (Gleskova, Morin, and Wagner 1993a,b) eight samples with the same properties were rapidly degraded until saturation was reached using light of intensity 3.4 W/cm². These samples were then annealed at any of four elevated temperatures, with two samples at each temperature, one annealed in the presence of light at 0.34 W/cm² and for comparison one annealed in the dark. Their results are shown in Figure 5.16; panel (a) has the dark relaxation data, panel (b) the data in light. It is clear that light significantly enhances annealing at all of these temperatures. Once this effect became recognized, it was applied in other ways; for example, it was found that the light-enhanced annealing properties of defects induced by light are different from those induced by unusual deposition of the material (Hata and Matsuda 1993).

Still another new effect was reported by Yang and Chen (1993a,b), who measured the response of *p–i–n* solar cells under various degradation and relaxation conditions. They reported two significant new results:

1. After a cell was rapidly degraded at an intensity of 50 suns and then allowed to stand under a 1-sun illumination, its relaxation transient at 1-sun intensity had a peak, as shown in Figure 5.17. That seems to be the first observation of a peak, which was not described by any of the existing rate equations. This effect was interpreted (Yang and Chen 1993a,b) as the consequence of the presence of two types of defects, one group with fast re-

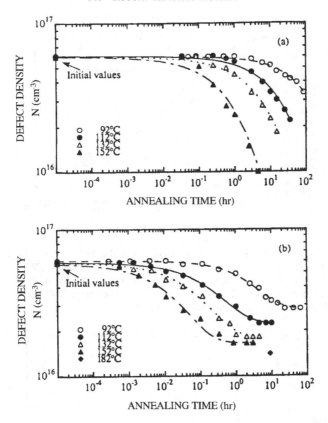

Figure 5.16 Light-enhanced annealing – (a) in the dark and (b) with 0.34 W cm^{-2} white light – at four temperatures. *Source:* Gleskova, Morin, and Wagner (1993a).

sponses, and one slow, although it is not clear if these simply represent different parts of a continuous distribution of time constants as mentioned earlier. Also, a completely different explanation of this effect has been offered, based on a continuous distribution whose optical and thermal dispersive properties are slightly different (Redfield and Bube 1993).

2. As shown in Figure 5.18, Yang and Chen (1993a,b) found that after light-induced degradation, the relaxation transients in the dark may depend on sample histories. They degraded two comparable cells to the same final reduced efficiency, but in one case the degradation was done rapidly by a 50-sun intensity, and in the other it was done slowly at a 1-sun intensity, with the durations adjusted to provide the same overall efficiency loss of the cells. The difference in the transient responses shown is important because it confirms the expectation that a single number – such as defect density or cell efficiency – is not a complete description of the material condi-

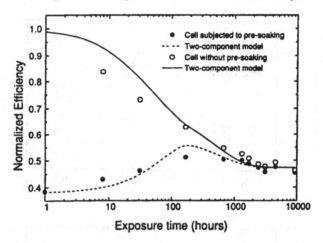

Figure 5.17 Normalized efficiencies of two identical a-Si:H *p–i–n* solar cells as a function of light exposure time under 1-sun illumination. One cell (●) was illuminated for 21 h at 50 suns prior to the 1-sun illumination, and the other (O) started from the as-deposited state. The solid and dashed curves are calculated from a two-component model. *Source:* Yang and Chen (1993a).

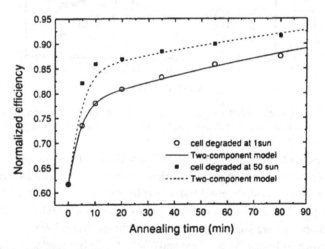

Figure 5.18 Normalized efficiencies as a function of annealing time at 130 °C for two a-Si:H *p–i–n* solar cells that were degraded to the same efficiency using different light intensities and durations. One cell (■) was illuminated at 50 suns for 5 min, and the other (O) at 1 sun for 100 h. *Source:* Yang and Chen (1993a).

tion. This is consistent with the picture of a distribution of values of the defect properties, since there might be changes in the *distribution* that affect the detailed properties of the material, but that might not be observable in the total number of defects. Such differences in behavior as in Figure 5.18 are not often seen, however.

These various complications in the kinetic properties of the metastable defects that could not be explained in previous models have led to several attempts to revise the earlier, simple pictures. We have already mentioned the proposal (Yang and Chen 1993a,b) that there are fast and slow groups of defects, and the alternative simulation by different optical and thermal dispersive parameters (Redfield and Bube 1993). To explain the possibly related light-enhanced annealing (Wu, Siefert, and Equer 1991; Meaudre and Meaudre 1992a,b; Graeff, Buhleier, and Stutzmann 1993), several variations of prior rate equations have been proposed, but these disagree with each other and have neither a reasonable physical rationale, a dispersive character, nor the symmetry that should exist in the equations. A newer proposal without these shortcomings is mentioned in the following section on models.

Two proposals that have symmetry and dispersion have been made, one to explain the peak in the transient in Figure 5.17, and another to explain light-enhanced annealing. For the peak, it was shown that if the optical generation and recovery terms in the four-term rate equation have a slightly different dispersive character from the thermal terms, the observed peak could be explained and well simulated (Redfield and Bube 1993). This requires dividing Eq. (5.4) into two parts, one with the optical terms and one with the thermal terms, and using two values of β. Light-enhanced annealing was shown to follow from the four-term rate equation if the coefficients are not constants, as had been assumed, but vary with either temperature or light intensity (Redfield and Bube 1994). Such a possibility had been mentioned earlier (Meaudre and Meaudre 1992a), but the close simulation (Redfield and Bube 1994) of the results given in Figure 5.16 using this approach provides a firmer basis. Nevertheless, none of these proposals can be regarded as confirmed as of this writing.

A still newer proposal to explain light-enhanced annealing has introduced an interesting variation on the dispersive rate equation in which the driving quantity for defect annealing is not the recombination (or capture) rate, but rather the carrier density (Meaudre, Vignoli, and Meaudre 1994). Although some other aspects of this analysis may be questioned (it does not specify a process by which carriers cause an annealing transition, and it is not symmetrical), the use of the carrier density explicitly in this way appears to offer promise more broadly. It seems also that the proposal given by Redfield and Bube (1994) could be directly recast into this terminology.

Also recently, some very unusual relaxations were reported in capacitance data that reflect the changes in charge state of previously charged defects as electrons are thermally emitted (Cohen, Leen, and Rasmussen 1992; Cohen, Leen, and Zhong 1993). Results suggest that an electron's thermal release rate is inversely proportional to its residence time in the defect; that is, as its time in the defect increases, its binding energy increases. Another unusual feature is

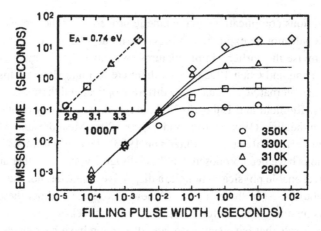

Figure 5.19 The dependence of the thermal emission time with filling-pulse duration for a sample that is lightly doped with phosphorus. The thin lines indicate
results of calculations based on non-Markovian processes. *Inset:* Variation of the
saturation values with temperature. *Source:* Cohen, Leen, and Rasmussen (1992).

the very long times (seconds at room temperature) over which these relaxations occur, as illustrated in Figure 5.19. Familiar rate equations were tried
and failed to explain this behavior (Cohen et al. 1992), and a "non-Markovian"
approach was found to provide a better fit; this involves a *memory* of the preexisting condition during relaxation. These reports have stimulated controversy
and at least two theoretical attempts to explain them in different ways (Branz
and Schiff 1993; Branz and Fedders 1994).

5.6 Microscopic Models of Photoinduced Defects
in a-Si:H

The ultimate goal of defect studies is, of course, to obtain a microscopic description of these defects, and to understand their origin and the mechanism by
which they form. Nearly all of this has been achieved in the case of the DX
center in AlGaAs (as described in Chapter 2), but progress in a-Si:H has been
limited. Although there have been proposals of microscopic models for the
dangling-bond formation, they often conflict and none is close to being established. There are many more proposals that attempt to model particular aspects
of defect properties or transitions without an atomic description of their underlying cause. Here, too, there is continuing disagreement in a number of areas,
even to the question of whether the defects are of intrinsic or extrinsic origin

(Redfield and Bube 1991a). Some models of defect kinetics have already been described in Sections 5.1 and 5.5 (e.g., the bimolecular-recombination and SE models) and are not repeated here.

For these reasons the following summaries of some of these modeling proposals are brief and intended only as guides to the literature. The preponderance of opinion is that the metastable defects in a-Si:H have intrinsic origins in the sense that they do not require the presence of any chemical or physical departure from a simple continuous random network; generally the ≈10 percent hydrogen needed in electronically good materials is considered to be an intrinsic part of the materials for this purpose. The evidence for an intrinsic origin was shown to be less than compelling, however (Redfield and Bube 1991a), and an extrinsic source was found to have significant support.

Adler's Model

The first detailed model was that of Adler (1981), based on successful modeling of properties of semiconducting chalcogenide glasses in which charged defects play a key role. The essence of this proposal was that in equilibrium there are always many Si dangling bonds, most of which exist in positive or negative charge states and so are not seen by ESR. Under optical excitation these capture photoexcited carriers, and thus convert to a neutral, rehybridized condition that is the paramagnetic metastable state. Thus carrier capture by the hypothesized charged defects is central to formation of the metastable state. This model deals only with the conversion by light of existing DB defects, not the formation of any new ones.

To provide the needed charged defects, Adler assumed that the defects have a negative effective correlation energy ($U < 0$) so that their neutral state would not occur in equilibrium. Many equilibrium results since then, however, have established the presence of spins on dangling bonds, which occur only on neutral defects; to explain this Adler added the hypothesis that inhomogeneous strains prevent some of the negative-U defects from relaxing into their lowest-energy, charged states (Adler 1983). More recently it has been pointed out that even if $U > 0$, charged defects are likely to exist in some regions because of local potential fluctuations, as indicated in Figure 5.20 (Branz and Silver 1990). One questionable feature of this diagram is the representation of E_C and E_V as flat in the presence of the potential fluctuations; they should vary for the same reasons that they follow the potential in a space-charge region.

This potential-fluctuation version of the Adler model was refined by Branz, Crandall, and Silver (1991), who cited much experimental evidence for charge trapping in the metastable transitions. They also pointed out weaknesses in

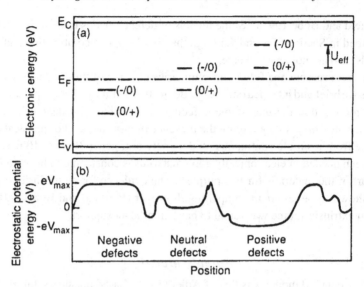

Figure 5.20 A random electrostatic potential (b), and its effects on the transition levels of the dangling-bond defect (a). In the present text, the flat band edges are questioned. *Source:* Branz and Silver (1990).

the more widely held alternative view that the dangling bond is formed by breaking *weak* Si–Si bonds (see next subsection).

One question in the Adler model that needs to be addressed further is why the density of defects in undoped material as measured by ESR agrees with the density measured by their optical absorption, which is not sensitive to the charge state. If there are more charged defects than neutral, the densities measured by absorption should be higher. Also, the densities of light-induced defects measured by the two methods agree.

Weak-Bond Breaking

The *weak-bond-breaking* model has become the most popular one for explaining formation of metastable defects. From the degradation of photoluminescence caused by light exposure (called "fatigue" by these authors), Pankove and Berkeyheiser (1980) first suggested that light-induced defects form by the breaking of weak Si–Si bonds, and the broken bonds provide nonradiative recombination paths for carriers. This emphasis on *weak bonds* has been widely used because it is known that the defects can be formed by excitation with light whose photon energy is as low as 1 eV (Fathallah 1990), although the Si–Si bond strength is about 2.2 eV. The degradation of optically detected magnetic resonance observed by Morigaki et al. (1980) was also attributed to formation of dangling-bond centers by light, and the increase or decrease in the strength of an ESR line at $g = 2.0055$ upon light exposure or annealing was interpreted by Dersch et al. (1981a) as the sign of the Si dangling bond. These were all consistent with the earlier identification of the built-in defects as dangling Si bonds (Brodsky and Title 1969).

For conversion of weak bonds to dangling bonds (by light or other excitation) there are many proposals that differ only in detail. One of the most specific was that of Stutzmann (1987), in which the transitions are intimately connected with charge-state effects. Like most weak-bond pictures, this one has a neutral ground state that converts to a dangling bond that may or may not may be charged, depending on the Fermi energy. Thus he points out that there are similarities between this kind of model and that of Adler, except that Adler favored a negative U and did not change the total number of DBs, only their charge state (Adler 1981; Branz et al. 1991). The related model by Stutzmann, Jackson, and Tsai (1985) for defect generation by electron–hole annihilation was discussed under kinetics in Section 5.1.

The Defect Pool Model

There have been numerous attempts to model data on metastable defects in terms of distributions of weak bonds, including those of Smith and Wagner (1987, 1989), Winer (1990, 1991), Powell and coworkers (Powell, van Berkel, and Deane 1991; Deane and Powell 1993b), and Schumm (1994). These have many common features:

1. All of them assume a broad distribution of possible energies for the dangling bonds within the energy gap – a distribution sometimes called a *defect pool*. (A similar distribution was used by Branz et al. [1991], but in the other defect-pool pictures it is assumed to be due only to structural disorder, not to random potentials.) No matter what their energies, all defects have the same value of U.

2. They associate each weak-bond state with a state of the valence-band tail and use the one-electron tail-state distribution as the distribution of initial energies for transitions to dangling-bond states. The valence-band-tail width is taken to be a measure of the structural disorder in the materials.

3. They use an equilibrium description of defect reactions as their starting point, and formulate the reactions in terms of a chemical-equilibrium equation relating the densities of weak bonds and dangling bonds.

4. Because of the broad distribution of possible defect energies hypothesized, charged as well as neutral defects occur for any value of E_F; these follow from the relations shown in Figure 4.3. In fact, some positively charged

defects are found in the upper half of the gap, as in Figure 5.20. It is the range of *local* environments that is responsible for the occurrence of defects with various charge states at different places.

5. Hydrogen atoms play a key role in *bond switching* to prevent the rejoining of a pair of dangling bonds that must result when a covalent bond breaks. There has been much effort to relate H to metastable defects; one widely cited study (Street, Hack, and Jackson 1988) concluded that the temperature dependence of the conductivity (which exhibits a kind of transition temperature) is attributable to the behavior of the set of H atoms acting as a glass. This *hydrogen-glass* model was intended to explain the freezing-in of some properties below a critical temperature, but the results were interpreted as being independent of changes in defect densities.

6. Because the energy to drive defect generation comes from recombining free carriers, their densities enter into the rates.

These analyses can be illustrated (apart from the role of carriers) by the following relation:

$$(Si\text{--}Si)_{wb} + Si\text{--}H \leftrightarrow [D^0 + (Si\text{--}H)] + D^0 \qquad (5.10)$$

where $(Si\text{--}Si)_{wb}$ is the initial, weak-bond (WB) state; Si–H is a *separate* complex of a Si atom that is bonded to one H atom; and D^0 is a neutral DB state formed when the WB breaks.

The essential feature is that the H atom moves from its initial site away from the WB and attaches to one of the two dangling bonds formed by the rupture, leaving one dangling bond at the former site of the H atom. This is the bond-switching process hypothesized to separate the two resulting D^0. The square brackets in Eq. (5.10) are intended to show the condition of the former WB after this transition to the defect state. To provide for the necessary H motion, diffusive processes of hydrogen are generally invoked (Kakalios et al. 1987). Treating this as an ordinary chemical-balance equation, a standard equilibrium relation for the densities of the various species is written that contains the usual thermally activated rate constant, which itself involves the free energy of formation of the defect. When this treatment is applied to the creation of charged defects, the value of E_F is introduced into the formation energy and thus into the equilibrium defect densities.

The defect-formation energy is taken to be the energy difference between that of the DB defect and that of the latent (WB) state, which is assumed to be a state of the valence-band tail. The density of states of the tail is obtained from the exponential optical absorption edge – the Urbach shape –

$$N_{V_t}(E) = N_{V_0} \exp(-E/E_U) \qquad (5.11)$$

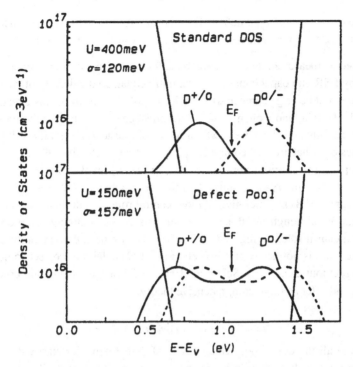

Figure 5.21 Calculated densities of states (DOS) in the gap, showing the difference between the "standard" two peaks (above) and the "defect-pool" DOS (below). Both calculations used the same defect-pool formula, the only differences being the parameter values displayed. *Source:* Schumm (1994).

with the deepest states thought to represent the weakest bonds. (Here the zero of energy is the value at which the exponential tail meets the more bandlike density of states.)

When the necessary distribution of defect energies is introduced to represent the varying local conditions that can affect the energy values, it is taken to have a Gaussian shape with a width σ. Then the resulting densities of states for the three charge states of the defect are calculated by a convolution of this Gaussian and $N_{V_t}(E)$. One representation of this density is shown in Figure 5.21, which compares the standard $N(E)$ with the defect-pool results for $E_F = E_V + 1.0$ eV. Both of these $N(E)$ arise from the same formula, the only differences being the assumed values of the effective correlation energy U and the Gaussian width σ. This illustrates some major results that are common to all versions of such calculations:

1. Departures from the standard double-peak $N(E)$ require that σ be at least comparable to U; this spreads defects with either charge state across the energy gap, as was shown in Figure 5.20.

2. The density of charged defects is typically much larger that the density of neutrals, even when E_F is near midgap, a new and unusual result.

More recent measurements, however, have found that the defect densities measured by ESR and capacitance are essentially equal, so the density of charged defects cannot be large compared to the density of neutrals in intrinsic material (Unold, Hautala, and Cohen 1994). Still other recent results give strong support to the value of $U = 0.3$ eV, which seems too large to permit the still larger values of σ that are needed in these pictures (Lee and Schiff 1992).

Comparing the various approaches, Winer (1990) invokes an analogy with Frenkel defects in crystals, rather than Schottky defects mentioned by Smith and Wagner (1987). Winer gives a more central role to hydrogen atoms in his formulation; although Smith and Wagner also invoke hydrogen, they do not give it a causal role, saying that such a role has not been convincingly demonstrated and is not necessary. Powell et al. (1991) claim to correct some errors in previous treatments. Schumm (1994) questions an assumption used by Winer, and offers a more comprehensive theory.

Non-Equilibrium Defect Densities

Although all the defect-pool calculations of defect densities start with equilibrium relationships, the desire for similar results for nonequilibrium (e.g., photoinduced) defects has led to application of these procedures to steady-state, nonequilibrium conditions (Winer 1990; Deane and Powell 1993b). To retain all of these relations it is necessary to assume that "the only effect of a nonequilibrium situation such as light soaking, carrier injection, or carrier depletion, is to change the free electron and hole concentrations and, as a result of that, the steady-state occupancy of the defects" (Schumm 1994). The quantity E_F is used even in these nonequilibrium conditions (Winer 1990).

One of the original motivations of these defect-pool analyses was the explanation of a set of data on the thresholds of subgap optical absorption by doped a-Si:H samples by Kocka (1987), who found that for n-type material the threshold is 1.1 eV, and for p-type it is 1.3 eV. These are generally thought to correspond to transitions from the (–/0) level to the conduction band, and from the valence band to the (0/+) level, respectively, and their sum is not consistent with the known energy gap of 1.75 eV and a positive value of U. It had been proposed by Bar-Yam, Adler, and Joannopoulos (1986) that a distribution of transition energies could produce such effects, and subsequent defect-pool treatments have been built on this proposal. This treatment has been used in various ways to explain other experimental results, including field-effect measurements in thin-film transistors (Deane and Powell 1993a).

Questions Relating to the Defect-Pool Model

Despite the many efforts to explain the data of Kocka (1987) on the basis of some sort of defect pool, it has been pointed out by Branz (1989), among others, that optical effects in localized centers should be treated in a configuration-coordinate (CC) framework rather than in a simple energy-level picture. Thus the transitions, which must be vertical in configuration space, *should* require energies greater than those represented by the levels (as detailed in Section 1.6). This removes a principal need for the defect-pool treatment (although it does not affect the principles), and it is not clear that other data require the kind of densities of states shown in the lower part of Figure 5.21.

In a different area, recent detailed studies by Ganguly and Matsuda (1993) show that the bulk defect density is determined by surface-controlled processes during material deposition, and not by bulk equilibration. Also recently, novel observations of relaxation of charged defect states have been interpreted as ruling out a defect-pool model (Cohen et al. 1993).

The role of hydrogen in defect-transitions is still uncertain. One problem is that photoinduced defect generation is now known to occur with about the same efficiency at temperatures down to 4.2 K (Stradins and Fritzsche 1994), thus excluding anything like diffusion processes from controlling the motion of hydrogen in defect formation. Alternatively, a very localized, nondiffusive role for H in bond breaking might be invoked, so that it becomes a part of the local configuration of the center in both its ground and metastable states. This conflicts, however, with the results of extensive magnetic resonance studies that demonstrate that (1) H atoms are separated by at least 5 Å from DB centers, and (2) no three-center bond with H occurs in the vicinity of a dangling bond (Tanaka 1991). These latter results agree with the accepted picture that H is preferentially located where it can relieve strain in an amorphous structure, and dangling bonds themselves relieve strain without the need for H. On the other hand, more recent measurements by pulsed NMR techniques found reversible changes in *local* H motion upon light exposure; one of several puzzling features of these results is that *most* of the 10^{21} cm^{-3} H atoms must change their motions even though only about 10^{16} cm^{-3} photoinduced defects are generated (Hari et al. 1994).

The hydrogen-glass model (Street et al. 1988) is difficult to relate to defect processes since the authors claimed that the defect density did not change in their experiments; however, there is considerable reason to believe that such changes must have occurred, and were in fact responsible for the observed effects. The similarity of *macroscopic* metastability, such as glass transitions, to localized metastable defects is too great for defect effects to be excluded (as was done by Street et al. 1988). There has been observation of limited photoinduced motion of H in some cases, but only well above room temperature where most measurements are done. An experimental test of the role of hydrogen was made recently by Vanecek et al. (1993), who reported that over the concentration range 0.29–12.6 percent, hydrogen does not affect the equilibration properties, in disagreement with the hydrogen-glass model.

The identification of valence band-tail states as weak bonds has played a central role in most of these calculations, but it is inconsistent with interpretations of comparable band tails in heavily doped crystalline semiconductors, in which there are no weak bonds (see the first seven papers in Landsberg [1985, 3–54]). Moreover, in those better-understood band tails the properties of bonds never play any role at all. Those band tails are spatial averages of quasi-localized states of mesoscopic scale produced by potential fluctuations. In a-Si:H the presence of strong potential fluctuations has been confirmed both experimentally (Hauschildt, Fuhs, and Mell 1982; Ley, Reichardt, and Johnson, 1982) and theoretically (Overhof 1992; Kemp and Silver 1993) without reference to bonds. Moreover, the evidence is overwhelming that the defect centers in a-Si:H are highly localized, so that a configuration-coordinate framework is appropriate to describe both their ground state and metastable state; therefore the initial and final states of a bond-breaking transition must be highly correlated spatially. This means that in Figure 5.20, for example, the energy of the ground state of a center should fluctuate locally with the potential as the DB states do, and cannot be taken (as shown there and as is generally done) as the spatially constant edge of the valence band or the constant energies of band tails. This correlation need was mentioned briefly (Smith and Wagner 1989), but has been overlooked in practices like convoluting the overall densities of initial and final states. In addition, optical transitions are always local, so the spatial correlation between initial and final states is complete.

The procedure of calculating nonequilibrium (photoinduced) effects by the use of equilibrium relationships with different values of carrier densities has not been justified, even though the carrier densities are coupled to the states of the defects. In the case of 1-sun light exposure the electron density increases by perhaps five orders of magnitude, so these are not small departures from equilibrium. This practice is equivalent to evaluating the nonequilibrium electron occupancy of deep levels in crystals by use of quasi-Fermi–Dirac statistics with a suitable quasi-Fermi energy as is done for shallow levels. That is known to produce erroneous results because the values of capture coefficients of the carriers are vital in establishing the nonequilibrium properties of deep levels. Moreover, even the normal dark relation for carrier densities $np = n_i^2$

(i = instrinsic) must be reconsidered in such cases, since these densities are coupled to the state of the defects (i.e., light exposure or annealing shifts the Fermi energy). For these cases the value of n_i depends on the extent of departure of the defect density from its equilibrium value, and may not be a fixed number whenever the defects are out of equilibrium.

Doping-Induced Defects

One of the early models for metastable defects was that of Street (1982b), who was concerned with the increase in the dark defect density caused by doping of a-Si:H, either *n*-type or *p*-type; Street proposed an explanation by the mechanism of self-compensation by native defects (described in Section 3.2). There are questions here in two areas, however. First is the assumption that doping is essentially different in a-Si:H than in crystalline Si because a-Si is overconstrained, so that phosphorus, for example, would not change from its favored threefold coordination to the fourfold in a-Si. The related assertion that a neutral P atom with fourfold coordination – designated P_4^0 – violates Mott's 8-N rule ignores the case in which the atom is neutralized by an orbiting electron, not a bonding electron. In the disordered material such an electron is likely to be in the band tail, and thus has nothing to do with bonds. So these arguments are oversimplified, and the actual energies of the P_3 and P_4^0 coordinations, both of which occur, must be determined before conclusions can be drawn.

The second area of question in the doping-induced-defect model for a-Si:H (Street 1982b) is the assumption that dopants produce self-compensating native defects as had been thought to occur in wide-gap crystalline semiconductors. However, as discussed in Section 3.2, that process seems not to be quantitatively applicable in important cases, and there is little reason to expect that it is more valid in a-Si:H than in those crystalline materials. In addition, an alternative explanation in crystals (the displaced-atom model described in Sections 2.4 and 2.6) is now well established, thus weakening the case for self-compensation by native defects. Equilibrium models also appear to be unable to describe low-temperature, light-induced defect formation.

The Rehybridized Two-Site (RTS) Model

This proposal is quite different from the others in that it invokes an extrinsic source for the metastable defects: either foreign atoms or physical defects (Redfield and Bube 1990; Redfield 1991). It was built on the successful explanation of the DX and related centers in III–V and II–VI compounds by a two-site model for a foreign atom (described in Sections 2.4 and 3.2). This choice was

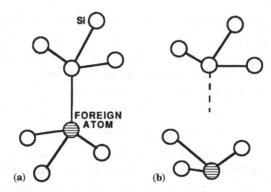

Figure 5.22 The proposed RTS model for the metastable center based on a foreign atom in analogy with the DX center in crystalline compounds: (a) ground state, (b) metastable state. *Source:* Redfield and Bube (1990).

made because of the many similarities in properties of the dangling-bond defect in a-Si:H and the DX center (Redfield and Bube 1990). In a-Si:H this model is most simply described in terms of a foreign atom in analogy with the DX center, as shown in Figure 5.22. In the ground state the foreign atom is in a substitutional site with fourfold coordination, and the center is electronically inert; in the metastable (defect) state one bond to a neighboring Si atom is broken, and the bonds of the foreign atom rehybridize to form three-fold coordination. A key feature is that although a foreign atom is involved, the defect state has a dangling bond on a neighboring Si atom as if it were an intrinsic effect; also, only one DB is produced, so there is no need for other bond switching.

This rehybridized two-site (RTS) model was shown to be consistent with many observed properties of the dangling bond (Redfield and Bube 1990; Redfield 1991). One prominent example was the finding that doping – either n-type or p-type – causes large increases in the density of defects induced by a given light exposure, as shown in Figure 5.23, in addition to increases in the density of built-in defects (Skumanich et al. 1985). The same authors found that the density of defects induced by light is even lower in compensated a-Si:H than in their best undoped material. These and other data led Skumanich et al. (1985) to the conclusions that

1. there is "a connection between dopant atoms, or doping-induced effects, and the light-induced defects," and
2. "creation of light-induced defects cannot simply be due to breaking weak Si–Si bonds."

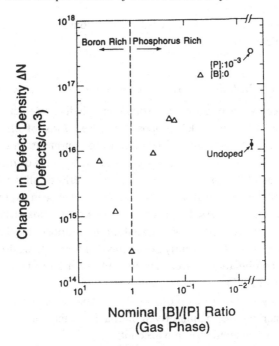

Figure 5.23 Dependence of the photoinduced defect density ΔN on the ratio of boron to phosphorus introduced into the gas phase during film growth. *Source:* Skumanich, Amer, and Jackson (1985).

One other example of a connection with foreign atoms was the finding of a metastable coordination change of added phosphorus atoms from fourfold to threefold induced by light (McCarthy and Reimer 1987).

Nevertheless, there are difficulties with the RTS model also. One is that the DX center that was used as its basis does not occur in crystalline Si, although it appears in many compounds. On the other hand, it is also true that there is no DB defect in crystalline Si like the one in a-Si:H. The most prominent difficulty with the RTS model is that dangling bonds occur in good undoped material, for which it is thought that there are too few foreign atoms to explain the $\approx 10^{17}$ cm^{-3} light-induced defects that typically arise there. Therefore it was suggested that the O, N, and C that are always present in sufficient densities might be involved, with C a good candidate because it easily forms both threefold and fourfold coordinations (Redfield 1991). Experimentally, the case for a role for C is ambiguous: Unold and Cohen (1991) reported an enhancement of light-induced degradation due to added C, but Morimoto et al. (1991) saw no change with added C. In a recent survey of possible effects of impurities, it was concluded that photoinduced defect densities are independent of impurity content for most impurities, with possible exceptions being O and C (Nakata, Wagner, and Peterson 1993).

Carrier-Induced Defect Formation and Annealing

Despite eighteen years of efforts to explain photoinduced defect formation in a-Si:H and the numerous models proposed in these efforts (of which those described here are not a complete list), there is no consensus on the nature of these defects, their origins, or a description of the kinetics of transitions between their ground and metastable states. In these efforts, all the familiar defects and processes in crystalline semiconductors have been explored as analogues, with limited success. It appears that these defects may have some unusual property that will need to be invoked for their further elucidation. Hence a different mechanism has recently been proposed for the transitions that does not rely on recombination of free carriers, but rather on just the *presence* of carriers (Redfield 1995).

There are several observed properties of metastable defects, not well explained by other, recombination-driven mechanisms, that this *carrier-induced* mechanism is designed to take into account:

1. Defects form in accumulation layers, where carrier recombination (or capture) is negligible.
2. Photoinduced formation of defects is as strong at 4 K as at 300 K.
3. The threshold photon energies for photoconductivity and defect formation are essentially equal.
4. Photoinduced defects shorten the carrier lifetime while they increase the extrinsic-to-intrinsic photoconductivity ratio.
5. Light enhances the annealing rate of excess defects.

The first of these observations had prompted previous suggestions for some carrier-induced mechanism, sometimes invoking thermal energy to drive defect formation (Schropp and Verwey 1987), but of course that could not apply at low temperatures. Also, none of these suggestions has shown *how* this could happen, or has related such a mechanism to photoinduced defect formation and to these other properties.

This new proposal is expressed in terms of the configuration-coordinate diagram in Figure 5.24, which has some similarities to that for the EL2 defect in GaAs (shown in Fig. 2.12) (Dabrowski and Scheffler 1989). For the first time in a-Si:H, it explicitly includes changes in the conduction-band edge induced by configuration changes, so that the conduction band (CB) is higher at a defect than elsewhere. Since the defects are the dominant recombination cen-

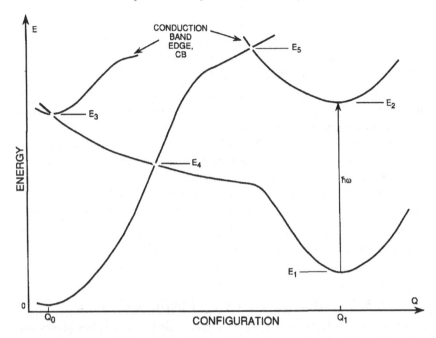

Figure 5.24 A possible configuration-coordinate diagram for the proposed carrier-induced defect-formation mechanism. The threshold phonon energy is $\hbar\omega$. *Source:* Redfield (1995).

ters, this may explain the long electron lifetime. An unusual feature of this CC diagram, though, is that at E_3 the latent center can convert to the defect state *with no energy barrier* by capture of a carrier from the CB.

The rest of this description assumes that the initial defect is negatively charged. In this picture photoinduced formation by photons of low energy starts by absorption of a photon *at an existing defect* (having configuration Q_1), thus explaining why the thresholds for photoconductivity and defect formation are equal and how an increase in the density of defects can shorten the carrier lifetime at the same time it increases the extrinsic-to-intrinsic photoconductivity ratio. The excited electron can move to a latent center (having configuration Q_0), thus possibly forming another defect. The new defect has the charge state of the initial one (assumed here as negative), whereas the initial one tends to restore its original charge state by capture of a free electron, thus lowering E_F. If such capture occurs, there are then two similar defects rather than one, and the process can proceed until it is slowed by the combination of the increasing rate of the inverse (photoinduced annealing) process and the limited supply of latent centers; a steady state eventually occurs. If electrons are not available for capture, the defect-formation process cannot continue. The lack of a barrier at E_3 explains the low-temperature formation, and the barrier $(E_5 - E_2)$ is that for light-enhanced annealing.

There are more facets of this proposal to be considered; one that is currently under investigation is the nature of the rate equation that is implied by such a mechanism. It is not yet clear if the unusual relaxations that have been reported recently by Cohen et al. (1992, 1993) lend themselves to description in these terms. The very slow relaxations in those observations are consequences of configuration changes that are induced by changes in the state of charge of the defects. Also, their dependence on the duration of an injection pulse may be associated with the property of charge-state effects that Kimerling (1978) has expressed as depending "on the length of time that the defect spends in the ... charge state." This is a provocative emerging area.

5.7 Photoconductivity Phenomena and Models

Dangling-bond defects in a-Si:H play a major role in reducing the photoconductivity of the material. A relationship of the opposite sense is also encountered in the common practice of measuring extrinsic photoconductivity to determine the density of such defects (e.g., in the constant photoconductivity method [CPM] technique, for a description of which see Section 4.5).

The dependence of photoconductivity on the type and density of defects, the Fermi energy E_F, the temperature, the doping, the transport processes, and the photoexcitation intensity are subjects of basic interest. Some of these effects are interrelated in a somewhat complex way. To introduce the subject, we first define the various possibilities.

Since photoconductivity in a-Si:H is predominantly due to electrons, we consider the issues from that perspective. The effectiveness of these defects as recombination centers for electrons depends on the density of available *empty* centers. It is necessary to consider, therefore, the variety of ways in which the density of available empty centers can be changed. Subsequently some specific examples are given.

1. Doping the material (without effects of illumination) increases the overall density of defect centers.
2. Doping the material (without effects of illumination) causes E_F to change, thus changing the fraction of empty centers.
3. Subjecting the material to light (at a fixed doping level) increases the overall density of defect centers.

4. Subjecting the material to light (at a fixed doping level) causes E_F to change as a result of the new defects formed by the light, thus changing the fraction of empty centers.

Leaving aside for the moment any changes in doping, it is clear, therefore that the formation of photoinduced defects affects the photoconductivity both through the increased density of defects present, and through the fraction of these defect centers that are empty due to changes in E_F caused by the change in the defect density. It is this "double" effect that must be kept in mind when interpreting results.

The effect of changes in E_F can be summarized as follows. If $E_F > E(-/0)$, there are few available electron capture sites until they are created by hole capture. As the Fermi level approaches and then moves below the $E(-/0)$ level, the density of available neutral defects increases, especially for low light intensities, and the photoconductivity decreases.

Several other effects related to photoconductivity are of general interest, which are discussed further later in this section:

1. A change in the dependence of photoconductivity on photoexcitation rate $\Delta\sigma \propto G^\gamma$, with γ decreasing from near unity to near one-half with increasing light intensity.
2. Examples are reported in which the photoconductivity exhibits the characteristics of superlinearity, thermal quenching, and optical quenching, usually associated with specific "sensitizing" imperfections in crystalline semiconductors.
3. Ordinarily one would expect approximate equality between the $(\mu\tau)_{ss}$ product obtained from steady-state photoconductivity, $(\mu\tau)_{ss} = \Delta\sigma/Gq$, and the $(\mu\tau)_{cc}$ product obtained from charge collection in "time-of-flight" drift measurements. This is because in a simple model, involving carriers of a single type, the drift mobility $\mu_{cc} = \mu_{ss}[n/(n + n_t)]$, where n is the density of free carriers and n_t is the density of trapped carriers, and the response time τ_{cc} is related to the lifetime τ_{ss} by $\tau_{cc} = \tau_{ss}[(n + n_t)/n]$, so that $(\mu\tau)_{cc} = (\mu\tau)_{ss}$. In the case of a-Si:H, however, $(\mu\tau)_{ss}$ is found to be about two orders of magnitude larger than $(\mu\tau)_{cc}$.

Effect of Doping

The solid symbols in Figure 5.25(a,b) show the dependence on E_F of the magnitude of the photoconductivity, and of the exponent γ in the power-law dependence of photoconductivity on excitation rate, according to the data of LeComber and Spear (1986). A high photoconductivity, relatively independent

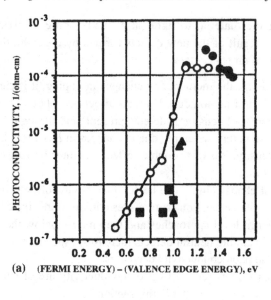

(a)   (FERMI ENERGY) – (VALENCE EDGE ENERGY), eV

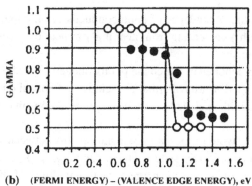

(b)   (FERMI ENERGY) – (VALENCE EDGE ENERGY), eV

Figure 5.25 Variation (a) of the photoconductivity in a-Si:H and (b) of the exponent γ (such that the photoconductivity $\Delta\sigma \propto G^\gamma$) as functions of the Fermi energy as varied by doping. The solid data points are from LeComber and Spear (1986); phosphorus doping (●), undoped material (▲), and boron-doped material (■). The open circles (○) and curves are the results for the simple model described in this section. *Source:* Bube and Redfield (1989b).

of E_F when the Fermi level is close to the conduction band, decreases sharply as E_F decreases below 1.0 eV, and becomes relatively independent of E_F again for lower values of E_F, three orders of magnitude smaller in the p-type material than in the n-type material. For E_F close to the conduction band, over the range in which the photoconductivity in Figure 5.25(a) is high, γ is about 0.55, increasing sharply to ≈ 0.90 when E_F drops below 1.0 eV.

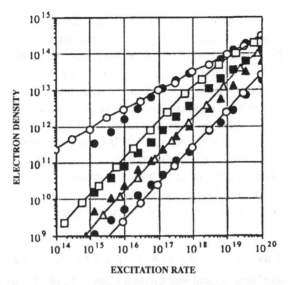

Figure 5.26 Photoexcited electron density (cm^{-3}) as a function of photoexcitation rate (cm^{-3} s^{-1}) for a sample of a-Si:H at four stages of optical degradation (corresponding to values of $E_F = 1.21$, 1.09, 1.04, and 1.00 eV, from top to bottom). The highest data are for an annealed sample, and the lowest data are for a degraded sample. Solid data points are inferred from the data of Staebler and Wronski (1980) using an electron mobility of 4 cm^2/V-s. Open points are for the curves calculated from the model of this section corresponding to the following values of E_F (eV), and assumed dangling-bond density (cm^{-3}): (upper ○) 1.21, 10^{16}; (□) 1.09, 6 × 10^{16}; (△) 1.04, 8 × 10^{16}; (lower ○) 1.00, 10^{17}. *Source:* Bube and Redfield (1989b).

Optical Degradation

Data on optical degradation available from Staebler and Wronski (1980) are given by the solid symbols in Figure 5.26. In the annealed state, E_F (calculated from the measured dark conductivity) lies 1.21 eV above the valence edge (somewhat higher than in undoped materials produced in subsequent years; half the band gap at room temperature is about 0.85 eV), the magnitude of the photoconductivity is large, and a value of $\gamma = 0.50$ is found for high photoexcitation rates. As optical degradation proceeds, the density of defects increases, the measured E_F decreases, the photoconductivity at a specific photoexcitation rate decreases, and the value of γ at high excitation rates increases to a value between 0.8 and 0.9.

Fermi-Level Shifts

Changes in E_F, whether due to doping, optical degradation, or annealing, can have major effects on the metastable defect occupancy and hence on the corresponding photoconductivity. These effects are of particular significance for the method of measuring defect density in terms of the changes in photoexcitation intensity needed to maintain a constant photoconductivity, the so-called CPM technique (described in some detail in Section 4.5) (Bube et al. 1992).

A related effect is the frequent observation during optical degradation or annealing of a-Si:H that the photoconductivity changes by a larger factor and/or with a different time dependence than the density of dangling-bond defects measured by CPM (Stradins and Fritzsche 1993; Bube et al. 1994). This would not be expected if the lifetime were simply inversely proportional to the defect density. It is often suggested that such effects are the result of more than one major type of recombination center; but at least in some cases tested in detail, simple accounting for changes in E_F is sufficient to describe the results.

Optical and Thermal Quenching of Photoconductivity

Phenomena involving optical and thermal quenching of photoconductivity, as well as superlinear photoconductivity over small temperature ranges, typical of the presence of sensitizing centers as well as recombination centers in semiconductors (Bube 1992), have been reported also for undoped a-Si:H (Vanier 1982, 1984; Vanier and Griffith 1982). Typical results are given in Figure 5.27. Such results have been interpreted to indicate that more than one type of recombination center may be active with different capture cross sections for electrons. Exposure of the material to high-intensity illumination produces a much larger effect in the shape and magnitude of the photoconductivity as a function of temperature at low temperatures than at room temperature.

A Simple Photoconductivity Model

In view of the fact that actual situations in a-Si:H may involve different kinds of defects acting as recombination centers, and various energy and capture-coefficient distributions of these defects as well, it is noteworthy that a simple model can give overall semiquantitative agreement with many experiments.

This model assumes that photoconductivity in a-Si:H is affected by only three types of localized states: the conduction-band-tail states, the valence-band-tail states, and localized states deep in the gap associated with dangling bonds, as shown in Figure 5.28 (Bube and Redfield 1989b). In the case of a-Si:H the defect associated with dangling bonds is considered to be a multivalent defect with three charge states. These three states are described as a negatively charged state with density [D⁻], occupied by two electrons; a neutral

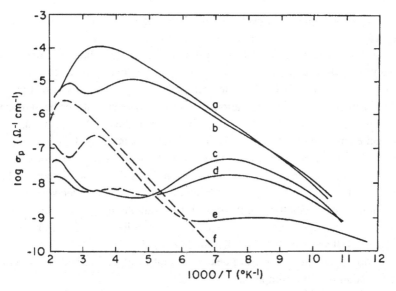

Figure 5.27 Temperature dependence of photoconductivity of six a-Si:H samples with different Fermi energies: $(E_c - E_F) = 0.31$ eV (a), 0.55 eV (b), 0.65 eV (c), 0.80 eV (d), 1.00 eV (e), and 1.29 eV (f). Solid lines represent photoconduction by electrons, dashed lines photoconduction by holes. *Source:* Vanier (1984).

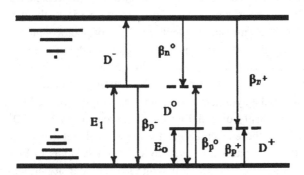

Figure 5.28 A simple model for photoconductivity in a-Si:H involving trivalent recombination centers with states D^-, D^0, and D^+, showing also the recombination coefficients β and the various transitions included in the model. Also included are an exponential distribution of conduction- and valence-edge tail states, the occupancy of which are calculated solely in terms of the location of the dark or quasi-Fermi levels. Unoccupied energy levels are shown dashed.

state with density $[D^0]$, occupied by one electron, which is therefore spin-active; and a positively charged state with density $[D^+]$, electron unoccupied. The energy E_1 in Figure 5.28 corresponds to the defect level $E(-/0)$, and the energy E_0 corresponds to the level $E(0/+)$.

The model includes exponential conduction-band-tail states but considers only capture transitions involving the recombination centers and electrons or holes in extended states. The density of trapped electrons or holes is calculated simply from the location of the dark or quasi-Fermi levels; the role of the tail states is therefore to behave as electron or hole traps. The model of Figure 5.28 also includes thermal excitation of electrons from the valence band to the D^+ center, thermal excitation from the D^- center to the conduction band, and thermal excitation of an electron from the valence band to the D^0 center. Specific values must be selected for E_1, E_0, and the various capture coefficients $\beta = Sv$, where S is the capture cross section and v is the carrier velocity. The choice of a discrete single value for E_1 and E_0 does not play a decisive role in determining the predictions of the model; for example, it has been shown that similar results are expected for a Gaussian distribution of levels (Bube et al. 1992), as well as for a variety of distributions of discrete levels.

We now compare the observed experimental results described above with the predictions of this simple model.

Doping

By combining data (LeComber and Spear 1986) on the dependence of photoconductivity on Fermi energy, and from Street (1985) on the dependence of defect density on Fermi energy, the dependence of photoconductivity was calculated as a function of Fermi energy using the model of Figure 5.28. In this calculation the density of dangling bonds was assumed to vary with doping from 10^{16} cm^{-3} for $E_F = 0.85$ eV to 1.3×10^{17} cm^{-3} for $E_F = 1.30$ eV in n-type material, and a symmetric behavior was simply assumed for p-type material with the density of dangling bonds increasing to 1.3×10^{17} cm^{-3} for $E_F = 0.40$ eV. For these calculations it was assumed that $E(-/0) = 0.90$ eV and $E(0/+) = 0.45$ eV. Results are given by the open symbols in Figure 5.25(a). Calculations show that the actual magnitude of the density of dangling bonds has much less effect on the photoconductivity than the value of E_F. There is a strong similarity, both qualitatively and quantitatively, between the predictions of the model and the data of LeComber and Spear. These results are particularly striking because, with the exception of the choice of $E(-/0) = 0.90$ eV and electron mobility $\mu_n = 10$ cm^2/V s (both close to reported values), no other parameters were adjusted to provide the fit. Even closer agreement with the experimental data of Figure 5.25(a) can be achieved by the choice of the two-defect model suggested by the analysis of dark-conductivity data in Section 4.6 and Figure 4.16.

Optical Degradation

Values of dangling-bond densities as a function of Fermi energy were taken from the data of Stutzmann et al. (1985) and Curtins et al. (1988) to correspond to Staebler and Wronski's (1980) data. The expected variation of photoconductivity with excitation rate was calculated for four samples in various stages of optical degradation. The results of these calculations over a restricted range of excitation rates are given by the open symbols in Figure 5.26.

In the annealed state, $E_F = 1.21$ eV (above the valence edge), the DB density is assumed to be 10^{16} cm^{-3}, and a value of $\gamma = 0.50$ is found over the whole range plotted. As optical degradation proceeds, E_F decreases toward $E(-/0)$ so that a significant fraction of the D$^-$ levels are converted to D^0 levels in the dark, and $\gamma = 1.0$ at lower light levels, for which the density of D^0 centers is essentially unchanged by the illumination.

The transition from $\gamma = 0.50$ to 1.0 shifts to higher excitation rates with increasing degradation. For the most degraded sample used, $E_F = 1.00$ eV above the valence edge, the density of dangling bonds is assumed to be 10^{17} cm^{-3}, and $\gamma = 1.0$ over this whole range.

The similarities between the curves of Figure 5.26 and the experimental data shown there from Staebler and Wronski (1980) indicate that the major effects can be described as the consequence of a decrease in the value of E_F caused by the increase in the total density of dangling bonds created by photoexcitation. Both the data and the calculations of Figure 5.26 indicate that the decrease in photoconductivity at lower excitation rates with optical degradation is greater by an order of magnitude than would be expected from a simple increase in the density of DB recombination centers.

Remaining to be explained is the cause of the discrepancy between the value of $\gamma = 1.0$ obtained from this simple model and the inferred values of γ shown in Figure 5.26. There is evidence that other recombination centers may need to be included in the model – for example, centers with small electron cross section (Vanier 1984), and the effects of pair-type recombination – as well as a distribution in energy and capture coefficient for recombination centers (Shepard et al. 1988). The description of γ has been the subject of several detailed investigations (Smith 1987; Almeriouh et al. 1991; Smail and Mohammed-Brahim 1993).

Fermi-Level Shifts

Use of this simple model indicates that appreciable errors can result if CPM is used to determine defect densities under conditions in which $E_F \leq E(-/0)$

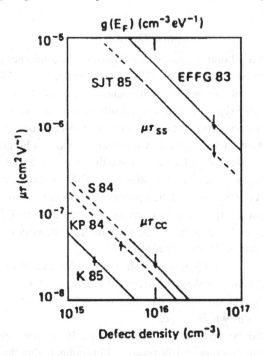

Figure 5.29 Comparison of the $(\mu\tau)_{ss}$ product deduced from steady-state photocon-ductivity and the product $(\mu\tau)_{cc}$ deduced from the time-of-flight charge collection for samples of a-Si:H with different defect densities. *Source:* Schiff (1987a).

(Bube et al. 1992). Since E_F commonly decreases with increasing degradation and defect density, the photoconductivity decreases more rapidly than the actual defect density, and the density is underestimated. The error is greater the great-er the value of R, the ratio of the capture coefficient associated with a Cou-lomb attractive capture to that associated with a neutral capture.

Incorporation of changes in E_F during optical degradation into the simple photoconductivity model also makes it possible to understand changes during optical degradation or annealing in which the photoconductivity shows a change by a larger factor than the defect density determined by CPM (Bube et al. 1994). If this model is used to correct the CPM densities, disagreements are resolved and the measured lifetime is found to be inversely proportional to the corrected defect density. Values for the electron mobility can be deduced by comparing results of the model with measured photoconductivity. Provided that the ratio of capture coefficients R is taken to be close to unity, it is found that this value of mobility is independent of time during degradation and lies in the previously reported range of 4–8 cm^2/V s (Street 1991b, p. 271).

Superlinear Photoconductivity

Examination of this simple photoconductivity model for a multivalent defect shows that there is the possibility for a range of superlinear photoconductivity in which the lifetime increases with photoexcitation intensity, provided that the ratio of capture coefficients $R \approx 1$ (Bube 1993). There is thus intrinsic to the model a mechanism for superlinear photoconductivity, thermal quenching, and optical quenching, without requiring a second set of different recombination centers as in the normal crystalline case. Where – or if – this occurs depends in a complex way on the value of E_F, the value of the capture-coefficient ratio R, the temperature, the actual defect density, and – if more complex distributions of defects are present – on the specific defect distribution.

One of the most striking dependences is that on the capture-coefficient ratio R. A large value of $R > 100$ prevents the depletion of D^0 levels essential for the observation of superlinearity. Because of this dependence of the effect on the magnitude of R, further investigation into superlinear photoconductivity may provide useful information about the magnitude of R in a-Si:H. The literature has several indications that R is of the order of unity, but a more detailed investigation indicated that R may be more of the order of 80–100 (Beck et al. 1993).

The $\mu\tau$ Problem

The "$\mu\tau$ problem" is pictured in Figure 5.29 after Schiff (Parker, Conrad, and Schiff 1986; Parker and Schiff 1986; Schiff 1987a,b): the mobility–lifetime product deduced from steady-state photoconductivity is about 100 times larger than the mobility–lifetime product deduced from time-of-flight charge collection.

To resolve this conflict, Kocka, Nebel, and Abel (1991):

1. eliminated anisotropy corresponding to the different geometries used for the two types of measurement;
2. showed that indeed $(\mu\tau)_{cc} = (\mu\tau)_{ss}$, provided that $(\mu\tau)_{cc}$ is deduced from time-of-flight primary photocurrent transients that are trap limited; and
3. concluded that hole trapping is the limiting process for recombination, so that as a result, the free-recombination lifetime of the majority electron is much larger than that of the minority hole.

Kocka et al. therefore suggested an explanation for the "$\mu\tau$ problem" based on the fact that $(\mu\tau)_{cc}$ deduced from time-of-flight measurements is the result of single-carrier, trap-limited transport, whereas $(\mu\tau)_{ss}$ deduced from steady-state secondary photoconductivity is a two-carrier, recombination-limited experiment. During the lifetime that controls the steady-state photocurrent (i.e., the free-electron-recombination lifetime), the electron is deep-trapped and reemitted many times. At low illumination levels $(\mu\tau)_{e,ss} \approx 100(\mu\tau)_{h,ss}$, mainly because hole trapping acts as a bottleneck for recombination, and because the ratio of the density of free electrons to that of free holes is high. From their work the authors deduced a correlation energy of 0.16–0.23 eV for the multivalent metastable defect in a-Si:H.

Other Photoconductivity Phenomena and Models

Since photoconductivity is an easily measurable quantity, it is natural that extensive measurements have been made of photoconductivity during optical degradation and thermal annealing, with the results of these measurements being used to obtain more detailed understanding of the photoconductivity process. Since a complete photoconductivity model will continue to be a subject of investigation, we report here briefly examples of the kinds of research that have been carried out.

Measurements of transient photoconductivity were used by Street (1983), a few years after the work on optical degradation of Staebler and Wronski (1977), to argue that the primary effects of optical degradation are to increase the density of DB defects, reduce the $(\mu\tau)_{cc}$ product for both electrons and holes, and leave unaffected the carrier drift mobilities. Thus he argued against two early alternative models that suggested optical degradation results either from a large increase in hole traps (and not from a change in the DB density) (Cohen et al. 1981), or from an electron trap with a very large barrier for both capture and emission (Crandall 1981).

There have been a variety of suggestions concerning the actual dominant recombination path and mechanism. Many of the suggestions made have been deemphasized in later research, and a generally accepted model can be discerned as developing, of which the model described above is a simplification.

Dersch et al. (1983) investigated the effects on photoconductivity induced by ESR. They found a resonant quenching of photoconductivity in samples with a high defect density, which they attributed to spin-dependent tunneling of localized band-tail electrons to singly occupied defect states. In materials with low defect density, they observed an additional quenching resonance that they identified with spin-dependent diffusion of localized band-tail holes to doubly occupied dangling bonds. They concluded that recombination occurs through trapped electrons and holes that recombine via defect states by tunneling and diffusion, rather than by capture of free carriers.

Guha and Hack (1985) concluded that a model involving recombination at dangling bonds alone is inadequate to describe experimental results of the dependence of photoconductivity on spin density, or the intensity and temperature dependence of photoconductivity. They further concluded that a density-of-states model with sharp exponential tail states due to disorder and a few well-defined peaks in the gap state distribution due to dangling bonds is inconsistent with experimental observations on space-charge-limited conduction. They favored a model in which room-temperature recombination is controlled by a continuous distribution of states located between the trap-quasi-Fermi levels.

McMahon and Xi (1986) measured photoconductivity as a function of temperature in the range 125–500 K. After optical degradation they attributed the decrease in photoconductivity to an increase in both DB recombination (high-capture-rate centers with a positive correlation energy near the Fermi level) at and above room temperature, and tail-state recombination at lower temperatures with an energy-dependent capture rate. In addition they attributed the decrease in photoconductivity below room temperature to light-induced donor defects with energy level 0.23 eV above the valence edge, where their presence did not cause an observable change in the Urbach-tail width.

In order to describe the temperature dependence of photoconductivity, Yoon, Jang, and Lee (1988) constructed a numerical model including recombination of free carriers through the exponential tail states and through a Gaussian distribution of dangling-bond states in terms of the Shockley–Read theory. They suggest that additional recombination of trapped electrons and holes in the tail states through the dangling bonds by tunneling needs to be included to describe low-temperature photoconductivity. Figure 5.30 shows typical variations of photoconductivity as a function of temperature compared with calculated curves incorporating this assumption. A similar model is described by Smail and Mohammed-Brahim (1991).

A detailed model for recombination at dangling bonds and band tails developed by Vaillant, Jousse, and Bruyere (1988) is essentially consistent with the models of McMahon and Xi (1986) and Yoon et al. (1988) described above. At low temperatures, the photoconductivity is determined by equilibration between the charge densities in the band tails, whereas at high temperatures it is controlled by the dangling bonds. Agreement with experimental data indicates that the capture cross sections of the band-tail states must be a function of energy, as suggested by McMahon and Xi.

Liu (1992) has described experimental results indicating that, above room temperature, the steady-state recombination lifetime of free electrons can be described by the free-electron quasi-Fermi energy alone, without the need for specifying carrier generation rate or temperature.

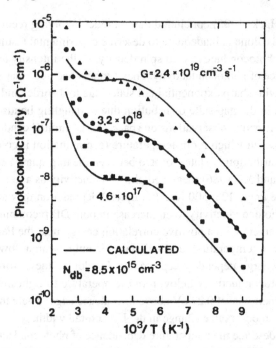

Figure 5.30 Experimental data for the dependence of photoconductivity on temperature for three photoexcitation rates. The curves are calculated according to the model of Yoon, Jang, and Lee (1988), including tail-to-dangling-bond recombination effects important at low temperature. The data are specifically for a dangling-bond density of 8.5×10^{15} cm^{-3}. *Source:* Yoon, Jang, and Lee (1988).

Estimates of Capture Cross Sections

In addition to the references given above citing values for R, the ratio of charged to neutral capture coefficients, there are a few other instances of determination of R and of the individual capture coefficients themselves. Street (1982a) measured the mobility, lifetime, and capture cross sections for the trapping of electrons and holes at DB defects using time-of-flight transient photoconductivity measurements. His results for neutral dangling bonds indicated a capture cross section of 4×10^{-15} cm^2 and a mobility of 0.5 cm^2/V-s for electrons, and a capture cross section of 2×10^{-15} cm^2 and a mobility of 3×10^{-3} cm^2/V-s for holes. These cross sections correspond to capture coefficients β_n^0 and β_p^0 of about 4×10^{-8} and 2×10^{-8} cm^{3}s^{-1}, respectively. Estimates of these values of capture coefficient by Pandya and Schiff (1985) from transient photocurrent methods are at least one order of magnitude smaller.

Recently, data obtained during optical degradation of undoped a-Si:H has been analyzed using the two-defect model described in Section 4.6 and Figure

4.16 (Bube, Benatar, and Bube, in press). The same model so successful in describing that dark conductivity is also successful in describing the photoconductivity and providing values for room-temperature capture cross sections: 1×10^{-16} cm^2 for the electron capture cross section of a neutral higher- or lower-lying defect, 2×10^{-16} cm^2 for that of a positively charged upper-lying defect, and 20×10^{-16} cm^2 for that of a positively charged lower-lying defect.

Dual-Beam Photoconductivity

On some occasions it has seemed advantageous to use photoexcitation by two different photon energies in order to unravel the photoconductivity process, in the same way that optical quenching of photoconductivity in standard semiconductors can be produced by simultaneous photoexcitation by a primary intrinsic and a secondary infrared light (Bube 1992, p. 76). In the case of a-Si:H, it is common to augment the usual photoconductivity measurement with an additional steady-state intrinsic background irradiation (the bias light), essentially adjusting the quasi-Fermi-level position by the background irradiation while measuring the photoconductivity due to the usual small-signal extrinsic test irradiation (the probe light).

On the basis of measurements of the spectral dependence of dual-beam modulated photoconductivity, Bullot et al. (1990) concluded that steady-state photoconductivity is due to at least three processes:

1. Electrons trapped on dangling bonds are thermally emitted into the conduction band.
2. Electrons trapped in the localized states of the conduction-band tail are thermally emitted into the conduction band.
3. Holes in the valence-band tail diffuse and recombine with D$^-$ dangling bonds.

In addition it is concluded that electrons in the conduction band are captured at D^0 defects.

Liu et al. (1993) measured CPM spectra under an intense bias light and found an anomalous, broad band in the energy region of defect absorption. They proposed that this band is caused by a combination of two processes:

1. excitation by the probe light from the valence band to the D$^+$ centers, increasing the recombination lifetime and enhancing the photocurrent; and
2. excitation by the probe light from the D^0 centers to the conduction band, decreasing the recombination lifetime and reducing the photocurrent.

The electron correlation energy of the defects was inferred to be 0.25 eV.

Modulated Photoconductivity Measurements

A technique for determining the energy distribution of energy states in the gap has been developed involving photoexcitation with light that varies periodically in time (Oheda 1981; Schumm, Nitsch, and Bauer 1988; Schumm and Bauer 1989; Kleider, Longeaud and Glodt 1991; Longeaud and Kleider 1992). The phase difference between the excitation light and the resulting photocurrent is measured as a function of the frequency of the modulated light, and the results are interpreted according to a model that yields a density-of-states distribution. It is found that behavior for high modulation frequencies is controlled by trapping and release of carriers, whereas the behavior for low modulation frequencies is controlled by recombination through deep levels. The technique has been applied to a variety of materials, including CdS (Oheda 1981) and a-Si:H (Schumm, Nitsch, and Bauer 1988; Schumm and Bauer 1989; Kleider, Longeaud and Glodt 1991; Longeaud and Kleider 1992). The results for a-Si:H indicate that the density of states determined in this way is for electron trapping, and that the major deep level lies about 0.6 eV below the conduction edge.

Particular Properties of Midgap States

Photoconductivity measurements have been used as part of a general investigation of the nature and effects of midgap states in undoped a-Si:H (Lee 1991; Gunes and Wronski 1992). It is suggested that results involving changes in electron $\mu\tau$ products, and sub-band-gap absorption and its dependence on generation rate, require two types of defects for effective modeling, one of which is different from the intrinsic dangling bonds. The new defects are located about 0.6 eV below the conduction edge, have higher electron capture cross sections by about a factor of 3 than the states above midgap after annealing, and are not detected by the usual CPM measurements. The effective density of both types of light-induced states above midgap increases about twice as fast as that of the states below midgap.

Effect of Surface Recombination

Modeling of the spectral reponse of photoconductivity in a-Si:H films for both the annealed and optically degraded states was used to indicate thickness-independent bulk properties for films with thickness between 1 and 3 μm, and surface and substrate interface recombination velocities of 3–6×10^4 cm/s and 1–2×10^6 cm/s, respectively (Li et al. 1991).

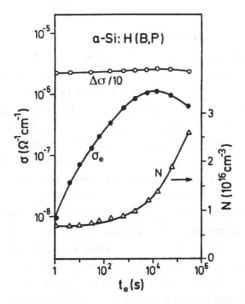

Figure 5.31 Dependence of photoconductivity $\Delta\sigma$, excess conductivity σ_e, and defect density N on exposure time t_e for a sample of a-Si:H doped with 30 ppm B_2H_6 and 150 ppm PH_3. *Source:* Rath et al. (1991).

5.8 Persistent Photoconductivity in a-Si:H

Persistent photoconductivity (PPC) has not been observed in homogeneous undoped a-Si:H, but has been reported in compensated, doping-modulated, and multilayer a-Si:H, and most recently also in singly doped a-$Si_{1-x}S_x$:H.

Compensated a-Si:H

Rath et al. (1991) reported measurements of photoinduced excess conductivity σ_e and defect density N on n-type a-Si:H films doped with 30 or 300 ppm B_2H_6 and various volume parts of PH_3. Figure 5.31 shows the time dependence under illumination of the photoconductivity $\Delta\sigma$, σ_e, and N at room temperature. Excess conductivity σ_e is ten times the dark conductivity σ_d after only 1 s of illumination. The measured defect density is similar to that found in undoped a-Si:H films with a similar E_F, and is attributed to the photoinduced formation of DB defects. Persistent photoconductivity apparently is very small or does not occur at all in p-type compensated a-Si:H (Mell and W. Beyer 1983).

Doping-Modulated a-Si:H

Doping-modulated a-Si:H – that is, alternating layers of n- and p-type doped a-Si:H – exhibits a room-temperature photoinduced excess conductivity, which has also been called persistent photoconductivity (Kakalios 1989). A few seconds of illumination suffices to increase the conductivity by several orders of magnitude, which decays with a time constant of days at room temperature, but can be removed by annealing at about 100 °C. Usually the dark conductivity of the n-type layers is many orders of magnitude larger than that of the p-type layers, and the excess conductivity is attributed to excess electrons that remain in the n-type layers after the end of illumination.

Hamed and Fritzsche (1989; also Hamed 1991) make a connection between the results for compensated a-Si:H and the doping-modulated a-Si:H by proposing that PPC effects in both have a common origin – namely, the existence of an aggregate of p- and n-type regions in the compensated material, approximating the situation in doping-modulated multilayers. In their model it is the normal introduction of dangling-bond defects in the p-type material by illumination that shifts E_F in the composite material toward the conduction band and produces the persistent conductivity without the introduction of any new or unusual types of defects; longer light exposure produces DB defects in the n-type material and once again acts to shift E_F toward midgap. It is reported that the experimental creation rate and annealing kinetics of the PPC in doping-modulated multilayers are correlated with the creation rate and annealing kinetics of the photoinduced conductance changes in isolated p- and n-type a-Si:H films with the same doping levels as the p- and n-type regions in the multilayer. Any effect of the p–n junction fields is believed to be minor.

a-Si:H Multilayers

This category includes the effects described in the last section under doping-modulated layers of a-Si:H. Indeed, Hamed (1991) argues that Hamed and Fritzsche's (1989) straightforward model of dangling-bond defect formation in the differently doped regions is adequate to describe the light-induced excess-conductivity effects in all kinds of layered structures, including n–i–n–i, p–i–p–i, and n–i–p–i multilayers.

Thermally Induced Metastability

As part of the present discussion of metastability in compensated and doping-modulated a-Si:H, it is appropriate to mention briefly analogous effects in

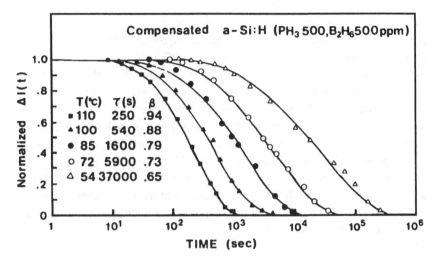

Figure 5.32 Relaxation of the normalized excess conductivity, induced by rapid cooling, as a function of time for different relaxation temperatures for a compensated a-Si:H sample. The curves are stretched-exponential fits to the data with the listed parameters. *Source:* Hyun et al. (1991).

thermally induced metastability in these types of material. Hyun et al. (1991) have investigated the conductivity changes induced by thermal quenching in compensated a-Si:H films with various compensation ratios of boron and phosphorus, as well as in doping-modulated multilayer structures. The magnitude of the excess conductivity upon rapid cooling (>10 °C/s) increases as E_F approaches the band gap. The excess conductivity relaxes slowly with kinetics that can be described by a stretched exponential as shown in Figure 5.32 for an exactly compensated sample, with values of $\beta = 0.65$ and $\tau = 3.7 \times 10^4$ s at 327 K, changing to $\beta = 0.94$ and $\tau = 250$ s at 383 K, corresponding to a value of thermal activation energy of 0.96 eV. It is suggested that diffusive hydrogen at high temperatures could activate dopants without changes in the DB density in compensated and doping-modulated films. The relaxation time in compensated samples is short compared to that in singly doped films because the barrier for hydrogen diffusion is smaller in the compensated materials, corresponding to the broader band tails in compensated a-Si:H.

a-$Si_{1-x}S_x$:H Doped Alloys

The first observation of persistent photoconductivity in a singly doped, single-layer a-Si:H sample was reported by Wang et al. (1993, 1994) for samples with modest n-type doping as the result of sulfur incorporation in the films.

Samples were prepared by plasma-enhanced chemical vapor deposition (PECVD) through decomposition of silane and hydrogen sulfide mixtures, and contained 3–6 percent sulfur. Observations of PPC were similar to those made for compensated and doping-modulated samples, as described above. For samples deposited at the same substrate temperature, the magnitude of the PPC at room temperature scales with the sulfur content in the films. The excess conductivity can be annealed at about 200 °C independent of the hydrogen concentration. The results in such singly doped materials suggest that the mechanism of charge separation due to internal electric fields may not be the complete explanation of the effect. It was suggested that the dependence of PPC in the a-Si$_{1-x}$S$_x$:H alloys on both the S and H content in the material indicates that defects associated with S and stabilized by H play a microscopic role in causing persistent photoconductivity.

6

Other Amorphous Semiconductors

In this chapter we consider examples of photoinduced defect processes in a number of other amorphous semiconductors for comparison with the results previously described. These examples are drawn from research on amorphous AlGaAs, compensated amorphous silicon, amorphous germanium, alloys of amorphous silicon and germanium, amorphous silicon nitride, and finally the amorphous chalcogenides. General references for these subject areas are *Disordered Semiconductors* (Kastner, Thomas, and Ovshinsky 1987), the *International Conference on Amorphous Semiconductors* (1993), and the review paper "Photoinduced effects and metastability in amorphous semiconductors and insulators" (Shimakawa, Kolobov, and Elliott, in press).

6.1 DX Centers in Amorphous AlGaAs Films

A comparison of DX-center effects in crystalline and amorphous Si-doped $Al_{0.34}Ga_{0.66}As$ has been reported by Lin, Dissanayake, and Jiang (1993). Two questions are treated: How are the relaxation properties of the DX center affected by changes from crystalline to amorphous? What is the connection between the DX type of defect in crystalline and amorphous semiconductors?

Below 250 K, DX centers in amorphous AlGaAs exhibit persistent photoconductivity behavior, such as is characteristic of their behavior in crystalline materials. The decay of PPC can be described by a stretched exponential with decay time constant τ and decay exponent β, such that τ decreases with increasing temperature. A comparison of the temperature dependence of τ for amorphous and crystalline material is shown in Figure 6.1, and a similar comparison of the temperature dependence of β is shown in Figure 6.2. The values of τ are larger for the crystalline than for the amorphous material, but the temperature dependence is approximately the same for both. The calculated barrier from the data of Figure 6.1 for electron capture is about 44 meV for the amor-

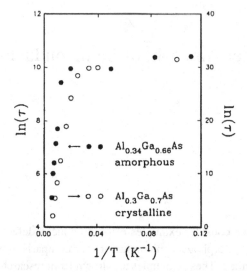

Figure 6.1 Stretched exponential time constant τ as a function of temperature for DX-center-related persistent photoconductivity decay for both amorphous ($\bullet$, left-hand vertical axis) and crystalline ($\bigcirc$, right-hand vertical axis) $Al_xGa_{1-x}As$:Si samples. *Source:* Reprinted from *Solid State Communications* 87, 787, Lin, Dissanayake, and Jiang, "DX Centers in $Al_{0.34}Ga_{0.66}As$ Amorphous Thin Films," © 1993, with kind permission from Elsevier Science Ltd., The Boulevard, Langford Lane, Kidlington OX5 1GB, UK.

Figure 6.2 Stretched-exponential exponent β as a function of temperature for DX-center-related persistent photoconductivity decay for both amorphous ($\bullet$) and crystalline ($\bigcirc$) $Al_xGa_{1-x}As$:Si samples. *Source:* Reprinted from *Solid State Communications* 87, 787, Lin, Dissanayake, and Jiang, "DX Centers in $Al_{0.34}$-$Ga_{0.66}As$ Amorphous Thin Films," © 1993, with kind permission from Elsevier Science Ltd., The Boulevard, Langford Lane, Kidlington OX5 1GB, UK.

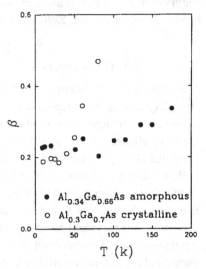

phous sample, which is about four times smaller than for the crystalline samples. The binding energy of DX centers in the amorphous material, determined from the temperature dependence of the dark carrier density, is about 100 meV, comparable to that of crystalline samples with the same composition. A combination of binding energy and measured thermal capture barrier gives a thermal emission energy (see Figure 2.4) of DX centers in amorphous AlGaAs of 144 meV.

Figure 6.2 shows that β increases almost linearly with temperature in the thermally activated capture region, but less steeply in amorphous than in crystalline, and is nearly independent of temperature in the low-temperature region. Another similarity is found in the magnitude of the Stokes shift, interpreted as the lattice-relaxation energy when a DX center captures an electron. A Stokes shift of about 0.9 eV was found in the amorphous material, compared to a value of 1 eV in the crystalline material. The similarity between Si-doped crystalline and amorphous AlGaAs suggests that a common description of DX centers may well be appropriate for both types of material (Redfield and Bube 1990).

6.2 Compensated a-Si:H

A number of differences are observed between the properties of undoped or singly doped a-Si:H and those of doubly doped, compensated a-Si:H (Street 1991b, pp. 158, 195). In compensated a-Si:H the band tails are broader, DB density and carrier mobility are smaller (Marshall, Street, and Thomson 1984), luminescence peak positions differ (Street, Biegelsen, and Knights 1981), and PPC is found similar to that in doping-modulated $n–p–n–p$ a-Si:H superlattices (Mell and Beyer 1983). Comparison of the drift mobility, extended-state mobility, and recombination lifetime in undoped and compensated a-Si:H indicate long-range potential fluctuations in compensated material as the cause of an observed decrease in extended-state mobility with increasing compensation (Tang et al. 1993).

A more detailed discussion of the occurrence of PPC and its interpretation in compensated a-Si:H is given in Section 5.8.

6.3 Amorphous Ge:H

Amorphous Ge:H has many similarities to a-Si:H, one of the most obvious differences being its smaller band gap of 1.1 eV compared to the 1.7-eV gap for a-Si:H. A review of research on a-Ge:H has been given by Paul (1991). Research on a-Ge:H has been limited (Karg, Boehm, and Pierz 1989; Martin et al. 1989; Turner et al. 1990), and only a few investigations have been made of metastability phenomena in a-Ge:H (Eberhardt, Heintze, and Bauer, 1991; Santos, Graeff, and Chambouleyron 1991). Ebersberger, Kruehler, and Fuhs (1993) have measured the dependence of the defect density on the Fermi energy for a series of differently doped a-Ge:H films. The defect-pool model (see Sec-

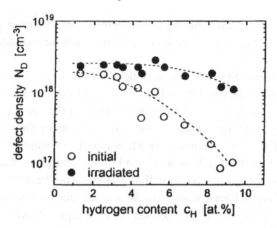

Figure 6.3 Defect density of a-Ge:H in the initial and the electron-beam-irradiated state as a function of the hydrogen content. *Source:* Scholz, Schroeder, and Oechsner (1993); © 1993 IEEE.

tion 5.6) was thought to be successful in describing the variation of defect density on Fermi energy. Extrapolation of the calculated defect densities gives a value of $3-6 \times 10^{16}$ cm^{-3} charged defects in undoped a-Ge:H, considerably larger than that in undoped a-Si:H, for which the corresponding value is in the range of 10^{15} cm^{-3}. On the basis of this defect-pool analysis giving the thermodynamic equilibrium values, no reduction in defect density in a-Ge:H below these values is possible.

Scholz, Schroeder, and Oechsner (1993) used 20-keV electron irradiation (see Section 5.3) to test the metastability properties of a variety of a-Ge:H films and found no substantial difference in stability behavior of a-Ge:H compared with a-Si:H. Their research showed that metastable defects can be created in a-Ge:H and that a saturated density can be measured. Typical results as a function of H content are given in Figure 6.3. The saturated density after electron-beam degradation was found to be almost independent of the initial film properties.

6.4 Amorphous Silicon Alloys

Amorphous SiGe:H alloys, with smaller band gaps than a-Si:H, and amorphous SiC:H alloys, with larger band gaps than a-Si:H, play an important role in solar cells based on a-Si:H and its alloys (see Section 7.4).

There is some evidence that a-SiGe:H alloys are less sensitive to light-induced degradation than a-Si:H (Nakamara, Sato, and Yukimoto 1983; Bennett et al. 1990), but the mechanism of light-induced degradation in a-SiGe:H

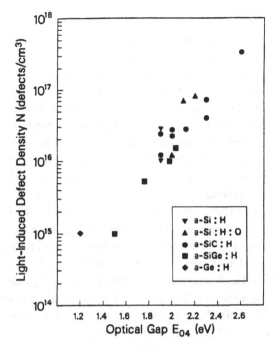

Figure 6.4 Upper limit values for the light-induced defect density N as a function of the optical gap energy for a-SiGe:H and a-Ge:H films. *Source:* Skumanich and Amer (1988).

alloys is perhaps even more debated than in a-Si:H itself. Skumanich and Amer (1988) have shown that there is a direct correlation between the optical gap and the photoinduced defect density in a variety of amorphous silicon alloys, the density of induced defects increasing by three orders of magnitude as the alloy gap increases from 1.5 to 2.6 eV, as shown in Figure 6.4.

Bauer, Schumm, and Abel (1993) have made a quantitative comparison of defect kinetics under high-intensity illumination on a-Si:H and a-SiGe:H, using a technique that produced the same generation rates and generation profiles in samples of different band gaps. Samples of a-Si$_{1-x}$Ge$_x$:H were used with band gap 1.45 eV $< E_g <$ 1.71 eV, corresponding to $0.42 > x > 0.0$. The $\mu\tau$ product in the undegraded state decreases with increasing value of x, and this is probably associated with a decrease in μ because of the observed broadening of the conduction-band tail with Ge alloying (Tanaka and Tsu 1981; Bauer et al. 1989). These latter authors invoke a model of chemical equilibrium (see Section 5.6) between weak bonds and dangling bonds (Schumm and Bauer 1991a,b), and derive saturated defect densities by introducing the deviation of the *np* product during illumination from the equilibrium product $n_0 p_0$ to describe changes of behavior with x and band gap. This is equivalent to lowering the defect formation energy by the amount $kT \ln(np/n_0 p_0)$.

Unold, Cohen, and Fortmann (1992) investigated the effects of illumination on a series of a-$Si_{1-x}Ge_x$:H samples $(0.25 < x < 1.0)$ using transient and steady-state junction capacitance with a semitransparent Pd Schottky barrier. The measurement technique allows discrimination between minority and majority carrier processes. The films were deposited using the photo-CVD technique, had narrow valence-band tails (50 meV), and reasonably low defect densities in the annealed state $(6 \times 10^{15}$ cm^{-3} for $x = 0.25$ to 1×10^{17} cm^{-3} for $x = 0.62)$. The $(\mu\tau)_p$ product for holes is considerably less (by a factor of 4) than its value in a-Si:H even for $x = 0.25$, and decreases further with illumination. The density of defect states deduced from optical spectra has two distinct components, above and below midgap, which increase with illumination; both are associated with Ge dangling bonds since no ESR evidence of Si dangling bonds was obtained in this investigation. The alloy films degrade similarly for all values of x, by a factor of 2–3 after 50 h of illumination at 400 mW/cm^2. Under these conditions, the $(\mu\tau)_p$ drops by a factor of 4–9, with the largest drop for the lowest band-gap alloy with $x = 0.6$.

6.5 Amorphous Silicon Nitride

It has been shown that ultraviolet radiation can generate an electron spin resonance (ESR) signal characteristic of silicon dangling bonds in Si- and N-rich hydrogenated amorphous silicon nitride (Kumeda, Yokomichi, and Shimizu 1984; Chaussat et al. 1985; Morimoto et al. 1985; Lowe, Powell, and Elliott 1986; Krick, Lenahan, and Kanicki 1987; Kanicki et al. 1991). The band gap of a-SiN_x:H is about 5.15 eV for $x = 1.6$ (corresponding to a Si-poor relative of the compound Si_3N_4).

Figure 6.5 shows the variation of light-induced spin densities with the composition of Si- and N-rich silicon nitrides. Only the ESR signal corresponding to silicon dangling bonds is observed in these films, indicating that nitrogen DBs are not induced under the experimental conditions used. The measured photoconductivity varies inversely proportional to the density of photoinduced silicon DBs, indicating that these defects are the dominant recombination centers in ultraviolet-illuminated, N-rich amorphous silicon nitride.

Measurements of (^{29}Si) hyperfine spectra and electron-nuclear double resonance (ENDOR) results (Lenahan et al. 1991) indicate that the photoinduced defect in N-rich and stoichiometric silicon nitrides is an unpaired spin, highly localized on a silicon atom back-bonded to three nitrogen atoms, which they

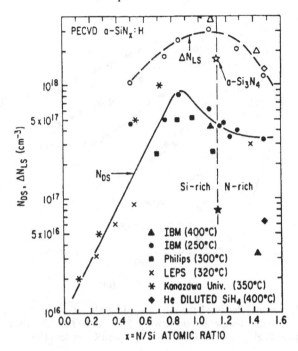

Figure 6.5 Dependence of the saturated value of the spin density before (N_{DS}, filled symbols) and after (ΔN_{LS} = [total spin – dark spin density], open symbols) broadband ultraviolet illumination of amorphous silicon nitride films as a function of the N/Si atomic ratio. Results from different laboratories (Lowe, Powell, and Elliott 1986; Chaussat et al. 1985; Morimoto et al. 1985) are given for comparison. *Source:* Kanicki et al. (1991).

called the *K center*. The K center has three charge states, K^+ ($N_3 \equiv Si^+$), K^0 ($N_3 \equiv Si^0$), and K^- ($N_3 \equiv Si^-$). The kinetics of formation of this K^0 (neutral paramagnetic state) center can be described by a stretched exponential, indicating the involvement of a dispersive process (Kanicki et al. 1990). The formation spectrum for photoinduced defects follows the absorption spectrum of silicon nitride, and a photon energy of at least 3.5 eV is required to create the K^0 center (Crowder, Tober, and Kanicki 1990; Kanicki et al. 1990).

The observation that the K^0 defect is bleachable (Crowder et al. 1990; Seager and Kanicki 1990) with light of photon energies between 1.8 and 4 eV, and that this photobleaching is temperature-independent and totally reversible, suggests that only an optically induced rearrangement of the charges in preexisting defects is involved, that is, the K centers are switched from a neutral paramagnetic state K^0 to a charged diamagnetic state (K^- and K^+) by photobleaching. This would simply reverse the initial process of K^0 formation by

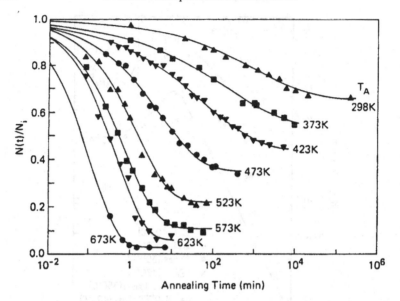

Figure 6.6 Isothermal decay of the normalized LESR signal intensity vs. annealing time for annealing temperatures between 298 and 673 K. The data are fit with stretched exponentials. *Source:* Tober, Kanicki, and Crowder (1991).

photoinduced emission of an electron from a K$^-$ state with subsequent trapping on a K$^+$ state.

Like the kinetics of formation of photoinduced defects, the kinetics of both thermal and photoinduced bleaching are also describable by stretched exponentials (Crowder et al. 1990; Tober, Kanicki, and Crowder 1991), as illustrated in Figures 6.6 and 6.7. From the thermal annealing as a function of annealing temperature shown in Figure 6.6, it can be deduced that the stretch parameter β increases approximately linearly with temperature for $T > 423$ K, and the SE time constant has an activation energy of about 0.43 eV (Jousse and Kanicki 1989).

Models involving bond breaking or hydrogen switching between N and Si atoms as induced by light seem to be ruled out by the observation that generation of defects is not accompanied by any change in the infrared (IR) spectra, in the etching rate, or in the thermal evolution of hydrogen from the film (Kanicki et al. 1991).

An analysis of data for the K defects in a-SiN$_{1.48}$:H indicates that the level corresponding to K(–/0) is located at about 3.5 eV below the conduction-band edge (Crowder et al. 1990; Kanicki et al. 1990), and the level corresponding to K(0/+) is located at about 2.6 eV below the conduction-band edge, giving a negative correlation energy, $U = [(E_c - E(-/0)) - (E_c - E(0/+))]$, of about -0.9 eV (Kanicki et al. 1991).

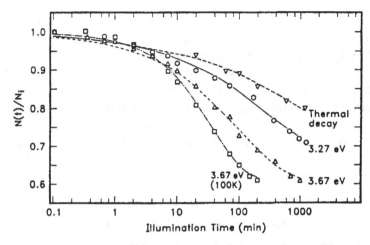

Figure 6.7 Time dependence of normalized LESR signal level for thermal annealing at 300 K, for photobleaching by 3.27- and 3.67-eV photons at 300 K, and for
photobleaching at 100 K with 3.67-eV photons. The curves are stretched exponentials. *Source:* Crowder, Tober, and Kanicki (1990).

In addition to the photoinduced creation of neutral K^0 centers with unpaired
spins, the photoinduced creation of fixed positive charges has also been observed in PECVD gate-quality N-rich silicon nitrides (Jousse and Kanicki
1989; Kanicki and Sankaran 1990; Kanicki et al. 1990). The kinetics of these
positive charges is also best described by a stretched exponential (Kanicki et
al. 1990), but the time constant and saturated value for positive charges are
smaller than the corresponding quantities for photoinduced spins. The positive
charges are located close to the nitride–silicon interface of metal–nitride–silicon structures. The photon energy threshold (Kanicki et al. 1990) is about the
same for the positive charges as for the light-induced spins, and the positive
charge can be photobleached with low photon energies (Kanicki and Sankaran
1990).

Ultraviolet illumination after thermal annealing at temperatures greater than
450 °C results in the creation of N centers, nitrogen dangling bonds (Warren,
Lenahan, and Curry 1990; Warren, Lenahan, and Kanicki 1991), with an energy level lying close to the valence-band maximum (Warren et al. 1991).

It was subsequently shown that ultraviolet illumination at low temperatures can also lead to the creation of neutral Si and N dangling bonds in N-
rich silicon nitride thin films (Warren, Robertson, and Kanicki 1993), indicating that as-deposited N-rich films contain both Si and N charged diamagnetic
centers. In a-$SiN_{1.6}$:H films, at 110 K, for example, only the K^0 center is initially created by illumination; after some time (420 s for the typical experimental conditions used) N_2^0 centers (neutral nitrogen–nitrogen bonds) appear,

but even at long times the increase in the K^0 density is greater than that for the $N_2{}^0$ density. It is suggested that two creation processes occur at the same time: charge conversion of K centers between K^- and K^+, $K^+ + K^- + h\nu \rightarrow 2K^0$, and charge transfer between $N_2{}^-$ and K^+, $K^+ + N_2{}^- + h\nu \rightarrow K^0 + N_2{}^0$. The $N_2{}^0$ center created at low temperatures has an identical ESR signature to that created by the annealing-plus-illumination procedure, but has a different annealing behavior: $N_2{}^0$ centers created at low temperature anneal away by 250 K, indicating why they have not been detected by room-temperature measurements; $N_2{}^0$ centers created by annealing plus illumination do not anneal until temperatures exceed 300 K.

6.6 Amorphous Chalcogenides

The amorphous semiconductors that contain Group VI atoms, usually sulfur, selenium, or tellurium, are commonly referred to as the *chalcogenide glasses*. The optical and magnetic properties of these materials have been reviewed by Taylor et al. (Taylor and Liu 1985; Taylor 1987a,b, 1989; Hautala, Ohlsen, and Taylor 1988; Ducharme, Hautala, and Taylor 1990; Liu and Taylor 1990). These amorphous semiconductors are usually made in bulk form by cooling from the liquid phase, and hence fall into the category of glasses. The most commonly researched materials in this category are Se, As_2Se_3, As_2S_3, $GeSe_2$, and GeS_2, and complex alloys.

Research indicates a number of metastable effects in the chalcogenide glasses involving defects with deep energy levels and metastable excited states. Essentially all these defects exhibit a negative effective correlation energy U. Typically the ground state of the defects are charged centers with energy level within 0.2 eV or so of the valence-band mobility edge, whereas the neutral, excited states of these defects have optical activation energies of about one-half of the gap, and thermal energies that are much smaller because of strong electron-lattice interactions.

Perhaps the most universally observed and striking optical effect in chalcogenide glasses is the so-called *photodarkening* (PD) effect. It consists of an increase in extrinsic absorption as the result of illumination with band-gap light, and is observed in almost all chalcogenide glasses (Tanaka and Kikuchi 1972; deNeufville 1976; Tanaka 1976, 1980b, 1984; Treacy, Taylor, and Klein 1979). Phenomenologically it is very similar to the increase in extrinsic absorption observed as a result of the illumination of hydrogenated amorphous silicon because of the effects of photoinduced defects. Other effects resulting from intense illumination of thin films include photostructural effects (deNeufville 1976) resulting in irreversible photopolymerization of the glass,

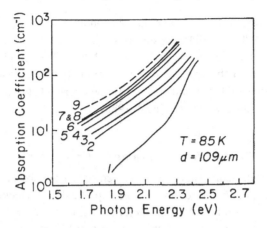

Figure 6.8 The effect of laser irradiation (150 mW cm^{-2}, 2.41 eV) on the optical absorption of glassy bulk As$_2$S$_3$ at 85 K with a sample thickness of 109 μm (the photodarkening effect). The numbered curves are as follows: (1) before laser irradiation, and after laser irradiation for a period of (2) 60 s, (3) 150 s, (4) 400 s, (5) 1,000 s, (6) 4,600 s, (7) 10,400 s, and (8) 17,500 s. Curves (7) and (8) are identical, indicating apparent saturation of the effect. Curve (9), however, was measured after an additional 13,200 s irradiation from the other side of the sample, indicating that the absorption coefficient varies appreciably across the thin sample. *Source:* Taylor in *Laser Spectroscopy of Solids II (Topics in Applied Physics,* vol. 65), ed. W. M. Yen, Springer–Verlag (1989).

and modulation of the index of refraction as a result of a thermal modulation of the band gap.

Photodarkening Effect

A typical example of the photodarkening effect is shown in Figure 6.8 in bulk glassy As$_2$S$_3$ (Liu and Taylor 1985). Irradiation of the sample with a laser at 2.41 eV, roughly equal to the band gap, causes the near-band-gap absorption to increase. The effect can be at least partially bleached by irradiation with photons of less than the band-gap energy (Tanaka 1976), and by thermal annealing near the glass transition temperature. A number of experimental measurements suggest a wide distribution of inducing and annealing times.

The PD effect appears to be the result of minor rearrangements of some electrons and atoms in the glass, rather than major, irreversible, structural changes. Because of the subtlety of the mechanism, there is still some uncertainty about the specific mechanism involved. There is general agreement that the effect requires the presence of Group VI atoms, in which the valence band consists of nonbonding *p*-electrons (Kastner, Adler, and Fritzsche 1975; Mott, Davis, and Street 1975; Ovshinsky 1976; Emin 1977; Jones, Thomas, and Phillips 1978; Ngai, Reinecke, and Enconomou 1978; Ngai and Taylor 1978; Tanaka 1980a,b, 1984). Possibilities suggested for relevant models include dangling chalcogen bonds (Biegelsen and Street 1980), undercoordinated and overcoordinated chalcogen atoms (Kastner et al. 1975), tunneling chalcogen atoms (Tanaka 1980b), and changes in the overlap between nonbonding chalcogen wave functions (Treacy et al. 1980).

Photoinduced Paramagnetic Centers

Four different metastable paramagnetic centers have been observed (Hautala et al. 1988) in the photoinduced (above band gap of 2.41 eV) electron spin resonance (LESR) spectrum of As_2S_3 at 20 K. Two of these (Type-I) centers, representing about 15 percent of all the photoinduced spins, anneal completely by 180 K. It has been suggested that the S_I center consists of a hole on a nonbonding $3p$ orbital of a sulfur atom, and that the As_I center consists of an electron on a nonbonding s–p-hydridized orbital on an arsenic atom. The other 85 percent of the photoinduced spins (Type-II centers) do not anneal completely until 300 K, and they have been attributed to a hole on a nonbonding $3p$ orbital of a sulfur atom (S_{II}), and an electron on a nonbonding $4p$ wave function on a twofold-coordinated arsenic atom (As_{II}).

There is not, however, a one-to-one correspondence between the PD effect described above and the LESR effect. Typical annealing curves for As_2S_3 are shown in Figure 6.9, comparing the LESR and the photodarkening. The difference in the annealing behaviors suggest that at least some of the states involved in photodarkening are diamagnetic. There is, however, a strong correlation in all measured properties between the optically induced increase in midgap absorption and the Type-I LESR centers. Confirming this are results that show that a 1 percent Cu alloy $Cu_1(As_{0.4}S_{0.6})_{99}$ displays no photodarkening but still shows metastable photoinduced ESR and midgap absorption (Liu and Taylor 1990).

Decrease of Photoconductivity with Time

Prolonged exposure of amorphous chalcogenides to light causes a decrease in photoconductivity. Many of these effects have been summarized by Shimakawa, Inami, and Elliott (1990, 1991). They are also in many ways similar to the effects observed as a result of the photoexcitation of a-Si:H.

A detailed investigation has been made of the kinetics of the decrease of photoconductivity under illumination in amorphous As_2S_3, As_2Se_3, GeS_2, and $GeSe_2$ films. The photocurrent I_p decreases with time of exposure and ap-

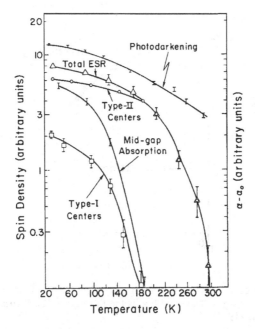

Figure 6.9 Comparison of the thermal annealing of the light-induced ESR, photo-darkening, and midgap absorption in glassy As_2S_3. The top and middle curves represent the thermal annealing of the change in average absorption coefficient $(\alpha - \alpha_0)$ at 2.41 and 1.30 eV, respectively, after optical excitation at 2.41 eV with about 250 mW cm^{-2} for 180 min at 30 K (right-hand scale). The data represent $(\alpha - \alpha_0)$ observed at 30 K after isochronal annealing for about 10 min at the indicated temperatures. The other three curves represent the density of optically induced ESR with the same conditions as for the PD. Total ESR intensity ($\triangle$); contributions of As_{II} and S_{II} centers ($\bigcirc$); contributions of As_I and S_I centers ($\square$). *Source:* Hautala, Ohlsen, and Taylor (1988).

proaches a constant value I_{sat} at long times. A typical example of the behavior is given in Figure 6.10 for As_2Se_3 where $\Delta I_p = I_p - I_{sat}$. The effects of pho-toexcitation can be removed by annealing near the glass transition tempera-ture. The four different metastable centers detectable by LESR experiments described in the preceding subsection (Hautala et al. 1988) all anneal below 300 K and hence do not participate in the effects shown in Figure 6.10.

It has also been shown that self-trapped excitons (STEs) (conjugate pairs of charged defects D^+–D^-) can be induced by illumination, but these would not be expected to affect the photoconductivity and anneal out near 300 K (Biegel-sen and Street 1980). It has been suggested (Shimakawa et al. 1990) that random pairs (RPs) of D^+–D^- defects might result from bond-switching reac-tions at STE centers, and that these could act as trapping or recombination

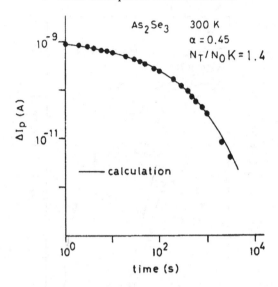

Figure 6.10 The decrease in photocurrent $\Delta I_p = I_p - I_{sat}$ as a function of time under illumination for a-As$_2$Se$_3$. The solid line is the fit to Eq. (6.2). *Source:* Shimakawa, Inami, and Elliott (1991).

centers, thus decreasing the photocurrent. The STE is produced from an excited, free-exciton state by illumination (Biegelsen and Street 1980).

The photocurrent is described by the expression

$$I_p = G/(N_0 + N_{RP}) \qquad (6.1)$$

where N_{RP} is the number of pairs of induced RP centers and N_0 is the original recombination center density. Assuming a dispersive form for the formation of N_{RP} – so that $N_{RP} = KN_T [1 - \exp(Ct^\alpha)]$ with $0 < \alpha < 1$, where N_T is the participating site density (Shimakawa et al. 1990) – gives

$$\Delta I_p = G/[(1 + KN_0/N_T) \exp(Ct^\alpha) - 1] \qquad (6.2)$$

which is the form of the curve drawn in Figure 6.10, with the values of the parameters indicated there. Similar results are obtained with a-GeS$_2$ with $\alpha = 0.50$ and $N_T/N_0 K = 0.7$. When measurements are made as a function of temperature, it is found that α increases with decreasing temperature – not the usual behavior for a dispersive system.

7

Photoinduced Defects in Devices

7.1 Introduction

The presence of metastable defects in device materials can have a variety of effects, some of them slight and some very important; but because the emphasis in this book is material phenomena, device effects are only summarized here. There are cases in which optically (or electronically) induced changes in defects degrade device performance, and some other cases in which similar processes speed the annealing of existing degradation. Either way, the inherent properties of metastability make its effects more prominent at lower temperatures, with the metastable-barrier energy determining the maximum temperature at which significant effects are observed. Since the energy gap of the host material limits the size of this barrier, Si devices operating at room temperature ($E_g = 1.1$ eV) are not much affected; GaAs ($E_g = 1.4$ eV) devices are affected, generally at lower temperatures; and compounds with larger gaps show the greatest effects. By far the greatest device impact is in photovoltaic cells of a-Si:H, whose degradation by bright light has been discussed at length in Chapters 4 and 5 and is the motivation for a great deal of the research reviewed there. There are effects in devices using crystalline III–V materials, but their impacts are less pervasive, and in other devices of a-Si:H they are not critical. For stable deep-level defects, the most prominent effect is as lifetime "killers" in minority-carrier devices, and indeed that is also the dominant effect of metastable defects in a-Si:H. In majority-carrier devices or semiconductor lasers other effects are more important.

One area in which effects were found quite early is in damage caused by ionizing radiation, which is of interest for devices operating in outer space or near nuclear reactors. The annealing of radiation damage in diodes of silicon (Kimerling 1975) or InP (Yamaguchi et al. 1984) is accelerated by forward bias. The interpretation of such effects as recombination-enhanced defect an-

189

nealing is now widely accepted. This process was described as relaxation of an atom from a nonequilibrium site at which it was placed by radiation damage; the released energy is from *local* modes of vibration that are excited by the conversion of electronic energy to vibrational energy as a carrier is captured at the site. In diodes, for example, a strong forward current can substantially increase the annealing rate by recombination of the many injected carriers as they are captured by recombination centers.

7.2 Devices Using III–V Compounds

Among the earliest observations of effects of deep-level defects in these devices were those in tunnel diodes, in which an *excess* current was often found. This excess current was traced to two sources: band tails in the heavily doped materials used, and defects with deep levels in the energy gap (Sze 1969, p. 171). If such a device were made out of an $Al_xGa_{1-x}As$ alloy with $x > 0.22$, then DX centers might increase the excess current. Most tunnel diodes, however, were made of GaAs, so the DX-center behavior did not appear.

A more prominent phenomenon in tunnel diodes was the increased degradation found at high operating currents (as happens also in *p–i–n* diodes of a-Si:H). By a series of measurements of this effect, Gold and Weisberg (1964) were able to attribute this degradation to newly forming point defects whose density scales with the nonradiative part of carrier-recombination energy during forward current flow. They proposed the term *phonon kick* for the effect that can help an atom move to a new position, thus creating vacancies or interstitials that can act as recombination centers. This motion is thus the same kind of effect as in annealing of radiation damage, although here it is harmful. The phonon-kick concept was the forerunner of *recombination-enhanced defect reactions,* which became widely accepted in later work on degradation of III–V lasers and light-emitting diodes (LEDs).

In early studies of those optoelectronic devices, it was found that their efficiency degraded rapidly when they operated at high currents or were exposed to intense light, both of which are found to produce an increasing density of dislocation networks. These networks, called *dark-line defects,* were evidently causing nonradiative recombination of injected carriers, as inferred from local reductions in brightness at the dislocations. It was then shown by Petroff and Hartman (1973) that the dislocations grew by an extrinsic climb mechanism involving either the absorption of interstitial atoms or the emission of vacancies. Hence atomic motions were taking place and could be considered as recombination-enhanced diffusion. With time, the materials improved enough to mitigate this effect before it was understood just how such motion could be driven by electronic processes providing at most 1.4 eV (the band gap of GaAs). These various effects have been reviewed by Lang (1982), who casts them all in a framework of recombination-enhanced reactions.

The DX center in $Al_xGa_{1-x}As$ alloys is known to have a deep level and is optically sensitive, as described in Chapter 2. Nevertheless, DX centers appear to have little effect on bipolar transistors or lasers made of these alloys, and it has therefore been concluded that these centers are not very effective recombination centers (Mooney 1990). It is also true that the active layer in most $Al_xGa_{1-x}As$ lasers contains very little Al, so the deep level of the DX center is not in the energy gap. A different effect occurs, however, in the wider-gap alloy layers that are heavily doped for conductivity purposes, and often have enough Al for the deep level of the DX center to be in the gap. In these layers donors are not as active as normal, and achieving the desired conductivity requires extra efforts.

The III–V device that seems most sensitive to DX centers is the modulation-doped field-effect transistor (MODFET), which has high-speed logic applications. That is because it depends on the action of a layer of $Al_xGa_{1-x}As$ in which $x \approx 0.35$, so the deep DX level is prominent. When electrons become trapped at low temperatures in the DX centers, they cause a shift in the threshold voltage of the device, a shift that is light-sensitive (Mooney 1990). In view of the current trend toward use of Ge–Si alloys for high-speed devices, however, this device, too, seems not to be coming into wide use.

The persistent photoconductivity associated with DX centers in materials such as AlGaAs:Si and CdZnTe:Cl can be used to write patterns of modulated carrier density in the material (Linke et al. 1994; MacDonald et al. 1994; Thio et al. 1994). This modulation in carrier density gives rise to a modulation of both the electrical conductivity and the dielectric constant, providing the capability for creation of an optical grating or hologram. In PPC, photoexcited carriers are free in the conduction band but spatially confined to the illuminated regions. It is estimated that a resolution better than 10^2 nm can be achieved. The observed change in the index of refraction, caused by the photoionization of the DX centers at 20 K together with the plasma effect, is thirty times larger than that observed in conventional photorefractive materials. It is persistent up to 35 K for AlGaAs and to 80 K for CdZnTe. The storage capacity of these materials is much larger than that of conventional photorefractive materials. Erasure of the pattern is readily achieved by heating above the annealing temperature of the PPC. Initial exploration of the effect was carried out with AlGaAs:Si, which has a PPC annealing temperature of only 50–100 K; but extension of the effect to II–VI materials carries the hope that room-temperature operation may become possible.

7.3 Devices Using II–VI Compounds

The degradation of Cu$_x$S/CdS solar cells, as described previously in Section 3.1, has a variety of different causes, and proved so difficult to control that this potentially useful solar-cell system has been abandoned. One of the mechanisms for degradation involves the diffusion of Cu from the Cu$_x$S into the CdS, producing a Cu-rich layer near the junction interface that responds to light by the formation of lifetime-reducing defects associated with the Cu.

The larger band gaps of II–VI compounds make them attractive for high-temperature electronic devices and for optoelectronic devices that could operate at shorter wavelengths. Existing LEDs and lasers made of III–V materials have been incapable of reliably producing green or blue light, and II–VI materials have long been studied to fill these needs. As described in Chapter 3, these efforts failed for many years because most of these materials cannot be sufficiently doped with both donors and acceptors to make the good *pn* junctions that are required for such diodes. The doping limitations, we recall, are due not only to solubility limits, but may also be due to any of several other causes. Prominent among these causes is the metastable, two-site configuration for the dopant atoms (like DX centers). The normal, substitutional site, which produces a shallow acceptor level, for example, acquires a higher energy (i.e., a metastable state) than an adjacent interstitial site that produces a deep level. The binding energy of the deep levels is too great for thermal excitation of carriers at room temperature, and inadequate conductivity results.

One target of special efforts for these applications has been ZnSe, which has an energy gap of 2.8 eV at room temperature, and can thus produce blue–green light. It can be readily doped *n*-type, but only slightly *p*-type, despite years of effort. Recently, however, a considerable increase in *p*-type doping up to 3.4×10^{17} cm^{-3} was achieved by use of a nitrogen-atom beam during growth in molecular-beam expitaxial process (Park et al. 1990b). This material was used to make an electroluminescent diode whose room-temperature emission spectrum is shown in Figure 7.1. This served as the basis for development of blue-green laser diodes ($\lambda = 511$ nm) with low threshold currents using alloys of MgZnSSe, ZnSSe, and CdZnSe for lattice matching (Haase et al. 1993). It has been found, however, that these diodes degrade rapidly during operation, rendering them insufficiently reliable for normal applications (Guha et al. 1993). It was shown (Guha et al. 1993) that the degradation is associated with the formation of dark-line defects fully comparable with those that had troubled early III–V devices. It is too early to tell now whether improved material-preparation techniques will succeed in suppressing this degradation.

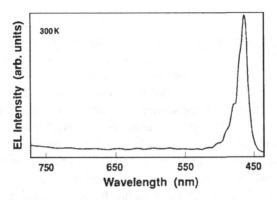

Figure 7.1 The electroluminescent spectrum of the first successful LED using nitro-gen doping for acceptors in ZnSe. *Source:* Park et al. (1990).

The success of *p*-type doping by nitrogen was not expected, despite its small covalent radius. Only by detailed calculations was it found that the N atom in ZnSe and MgSe has its lowest energy in the normal substitutional site that produces a shallow level. Attempts to achieve similar doping by faster metalorganic processes have been less successful, apparently because hydro-gen, introduced at the same time, partially passivates the acceptors.

7.4 Solar Cells Using Amorphous Silicon

The promise of hydrogenated amorphous silicon for low-cost photovoltaic power generation has stimulated many efforts around the world to develop such devices. This development has met with many successes in higher effi-ciency, uniformity, and reproducibility; large-size modules; fabrication of stacked junctions to exploit the solar spectrum more effectively; and cost re-duction. Nevertheless, the widespread use of such devices is being severely limited by the instability in the material caused by photoinduced metastable defects. Most solar cells of a-Si:H in commercial use lose about 30 percent of their generating capacity due to this effect alone, and must be "derated" ac-cordingly. The extensive research programs reviewed in earlier chapters were stimulated more by this device degradation than by the scientific interest in elucidating the behavior. The fact remains, however, that, although the first report of this Staebler–Wronski (1977) effect (shown in Figure 1.2) appeared close to twenty years ago, the understanding of the nature or origin of the ef-fect is still limited.

The initial work on this instability was on material properties, but it was then shown that the conversion efficiency of *p–i–n* solar cells degraded in analogous ways under light exposure (Staebler et al. 1981). From then on, research on homogeneous films and device structures became inextricably linked, with useful information obtained in each realm and applied to the other. Two of the most important observations that guide understanding of defect formation came from solar-cell studies (Staebler et al. 1981). First was the finding that photoinduced degradation of solar cells is lower when a cell is short-circuited rather than open-circuited while exposed to light, and can be prevented simply by application of a reverse bias to the device during light exposure. Since light absorption in the material or device is the same regardless of bias, this led to the conclusion that defect formation is not caused directly by optical absorption, but must involve photoexcited carriers, which are swept out by a sufficient reverse bias. Supporting this interpretation was the additional observation (Staebler et al. 1981) that *in the dark* these devices degrade similarly as a result of forward current flow, which, of course, increases the carrier density in the i-layer of the device. Together, these led to the general belief that metastable defects are formed from latent centers by taking up the energy released during capture or recombination of photoexcited carriers. This belief lasted until the reports of defect formation in accumulation layers of a-Si:H (discussed in Section 7.5) showed that defects can form in the absence of carrier capture or recombination. This affected the proposed microscopic models as summarized at the end of Section 5.6.

Among other observations linking homogeneous films and devices was confirmation of the expectation of the importance of carrier sweep-out by reports that degradation is lower in cells with thinner i-layers. One of the most comprehensive studies of solar-cell degradation was that of Chen and Yang (1991), who studied the dependence of degradation on both light intensity and temperature. Their measured results are shown in Figure 7.2 as points, and their fits of the data to theoretical kinetics formulas as the associated curves. An interesting aspect of that work is that the curves used so successfully there are the stretched exponentials that were developed (Bube and Redfield 1989a; Redfield and Bube 1989) to describe just the defect densities in undoped, homogeneous films (see Eq. [5.5]). The values of the parameters used for these cells are also essentially the same as for films. This is only one of many indicators that the light-induced degradation of device efficiency in a-Si:H solar cells is dominated by an increase in density of metastable defects in the i-layer of the devices, thus shortening the carrier lifetime. This conclusion is occasionally disputed because of the possibility that interfacial effects at the p–i or i–n boundaries may also contribute to degradation; this possibility cannot be completely ruled out, but it has been shown to be quite limited in effect (Redfield and Bube 1991c). For recent events in these areas, see the

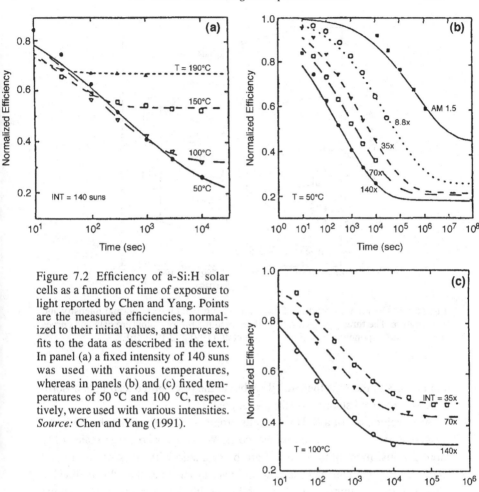

Figure 7.2 Efficiency of a-Si:H solar cells as a function of time of exposure to light reported by Chen and Yang. Points are the measured efficiencies, normalized to their initial values, and curves are fits to the data as described in the text. In panel (a) a fixed intensity of 140 suns was used with various temperatures, whereas in panels (b) and (c) fixed temperatures of 50 °C and 100 °C, respectively, were used with various intensities. *Source:* Chen and Yang (1991).

conference proceedings of the Materials Research Society (1994) and the *International Conference on Amorphous Semiconductors* (1993, esp. p. 323).

The highest data set in Figure 7.2(b) shows the solar-cell behavior at approximately normal intensity and temperature, and the degradation exceeds 50 percent after a long time. This severe loss is tempered in normal practice by two factors: First, in normal operation the light is not continuous, so some annealing can occur; second, thinner *i*-layers that are increasingly used degrade less. This latter point also supports the above-stated belief that defect formation requires a role for carriers, not just the absorption of light. In recent a-Si:H solar cells it has become common practice to reduce the thickness of

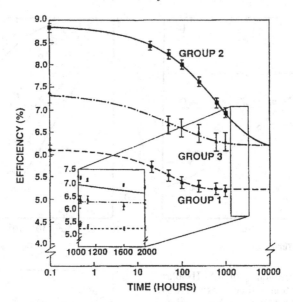

Figure 7.3 Degradation studies of large modules in field tests at normal intensity and temperature. The three groups are from three manufacturers; points are data and curves are stretched-exponential fits. *Source:* Luft et al. (1991); © 1991 IEEE.

the *i*-layers to increase their stability, but at the price of reduced light absorption, so improved stability is still a vital goal.

As development of a-Si:H solar cells progressed, their size has been steadily increased to reduce the unit-power costs. With the goal of large-scale utility applications, monolithic modules are being made with sizes of 30 cm and more. These modules have somewhat lower efficiencies than do small cells, but they are still improving. The stability problem in these modules is quite comparable to that in small cells, as can be seen in Figure 7.3. The data points are from field tests performed at the National Renewable Energy Laboratory on three groups of large modules from different manufacturers (Luft et al. 1991). The curves are again stretched exponentials, which were found to provide some predictive value once the curve shape was established. Thus SEs can describe the defect formation and annealing in homogeneous films, and the degradation kinetics of the efficiency of small solar cells or large modules. Despite all this, a microscopic description of photoinduced defect formation is still debated.

A parallel development has been the use of alloys of a-Si:H with germanium for lower energy gaps, or with carbon for higher ones (see Section 6.4). By a series stack of two or three *p–i–n* junctions with different gaps, in-

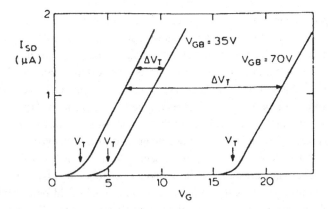

Figure 7.4 Examples of transfer characteristics of a thin-film transistor before (*left*) and after (*right*) bias stress. I_{SD} is the source-drain current, V_G the gate voltage, V_{GB} the previously applied gate bias (removed before these measurements), and V_T the threshold voltage. *Source:* Powell et al. (1987).

creased use of the solar spectrum is expected to furnish higher efficiencies. One incidental gain from this design is that each elemental i-layer is thinner than it would be by itself, since it is not expected to absorb the full spectrum, and so its degradation is reduced. However, these stacked cells have a number of problems of their own, one of which is that a-SiC:H is much less stable than a-Si:H. There has been some conjecture that this is a direct and unavoidable consequence of a larger gap, but the data do not yet firmly support this inference. For a-SiGe:H alloys there have been reports of better stability, but it has also been claimed that there is a lack of correlation between defect density in the bulk and stability of its devices (Xu, Yang, and Guha 1993). On the other hand, detailed studies of combined variations of the content of H, Ge, and Si in alloy *films* have been used to make what is claimed to be the most stable a-SiGe:H solar cell (Terakawa et al. 1994).

7.5 Thin-Film Transistors Using Amorphous Silicon

Thin-film transistors (TFTs) – field-effect devices first studied in Walter Spear's group (Snell et al. 1981) – are now generally used as the pixel drivers in active-matrix liquid-crystal displays. It was soon found that they possessed an instability when stressed with light or applied gate bias; this appeared as a shift in the threshold voltage of the device (Powell and Nicholls 1983). A simple version of this shift is illustrated in Figure 7.4 from Powell et al. (1987), showing the transfer characteristic curves (the source-drain current vs. the applied gate voltage) following application of a strong bias voltage to the gate. It is clear that such degradation can reduce the usefulness of the transistor, since increasing the threshold voltage reduces the current.

Two different explanations of these shifts were offered: one invoking the formation of charged metastable defects in the semiconductor itself, and the other invoking the presence of charges trapped in the insulator layer between the gate metal and the a-Si:H. Both of these processes are known to occur under some conditions, and extensive debate ensued. A persuasive set of elegant experiments by Hepburn et al. (1986) seems to have established that both types of effects are present. Some dispute over this conclusion remained, however, with arguments that all of the observations could be attributed to trapped charges in the insulator (Gelatos and Kanicki 1990). This appears to be a minority view, and the field-induced formation of metastable defects is widely accepted.

One important consequence of the formation of metastable defects in an *n*-type accumulation layer near the gate is that, since there are effectively no holes, there is negligible recombination. This raises questions for most models of defect generation, which depend on recombination energy to initiate the effect, and it has implications for microscopic models of these defects that are still being studied. One recently proposed (Redfield 1995) mechanism for defect formation that avoids this problem was summarized at the end of Section 5.6.

Apart from the interpretation of these instabilities, they appear not to create much of a problem for ordinary TFTs. In a recent two-volume book on devices of amorphous and microcrystalline semiconductors, there is little mention of these effects (Kanicki 1992, esp. vol. II, pp. 428, 462).

7.6 Xerography

During a xerographic cycle, the surface of a photoconducting film is first uniformly charged to a high voltage, then exposed to a patterned light flash that discharges the film in the exposed regions. If deep trapping of photocarriers occurs, it forms a remnant fixed charge that is equivalent to a residual potential V_R, and degrades subsequent xerographic performance. It has been found that exposure to light enhances such deep trapping in a-Se and in a wide range of binary and ternary glassy films of As:Se:Te compositions (Abkowitz 1987). Figure 7.5 shows this photosensitization of trapping of holes in a-Se caused by various durations of exposure to white light (at zero field) prior to measurement. For a 15-min exposure, for example, the effect is to increase V_R by a factor of 10 over that of the dark rested sample.

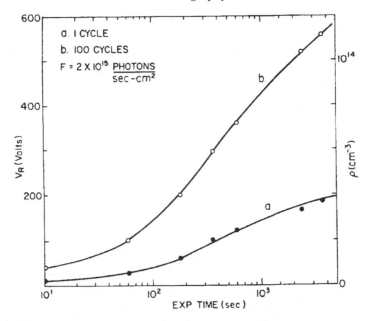

Figure 7.5   Residual voltage of a xerographic cycle vs. exposure time to light in a film of a-Se. The increase in V_R is an effective degradation. *Source:* Abkowitz (1987).

Although a full explanation of this degradation has not yet been given, several properties have been determined (Abkowitz 1987). By a xerographic time-of-flight technique, the profile of trapped holes was measured and found to agree with the distribution of absorbed photons from the 3,300-K radiant source used. A metastable character of these deep traps was established by re-laxation studies that found that the persistence of the traps greatly exceeded the carrier release times. It is thus concluded that trapping of a carrier on a metastable site does not cause immediate decay of the trap. In this respect these metastable defects are similar to those of the dangling bonds in a-Si:H, in which the defects can have any of three states of charge.

Analogous behavior was observed for photosensitization of electron trap-ping (Abkowitz 1987). Analyses showed also that the distributions of meta-stable traps for both electrons and holes after photosensitization is the same as before, so that only the integrated number of each (native) trap species ap-pears to increase during exposure. It is also found that the mobility–lifetime product of the carriers varies inversely with the number of deep traps, where-as the drift mobility, which is controlled by shallow traps, is not influenced by the traps that control the mobility–lifetime product.

References

Abkowitz, M., in *Disordered Semiconductors,* ed. M. A. Kastner, G. A. Thomas, and S. R. Ovshinsky (Plenum Press, New York, 1987), p. 205.

Adler, D., *J. Physique (Paris)* 42, C4-3 (1981).

Adler, D., *Solar Cells* 9, 133 (1983).

Adler, D., in *Semiconductors and Semimetals,* vol. 21, ed. J. I. Pankove (Academic Press, Orlando, 1984), pt. A, p. 291.

Almeriouh, Y., J. Bullot, P. Cordier, M. Gauthier, and G. Mawawa, *Philos. Mag. B* 63, 1015 (1991).

Amer, N., and W. Jackson, in *Semiconductors and Semimetals,* vol. 21, ed. J. I. Pankove (Academic Press, Orlando, 1984), pt. B, p. 83.

Austin, I., T. Nashashibi, T. Searle, P. LeComber, and W. Spear, *J. Non-Cryst. Solids* 32, 373 (1979).

Baba, T., M. Mizuta, T. Fujisawa, J. Yoshino, and H. Kukimoto, *Jpn. J. Appl. Phys.* 28, L891 (1989).

Baj, M., L. Dmowski, M. Konczykowski, and S. Porowski, *Phys. Status Solidi A* 33, 421 (1976).

Bar-Yam, Y., D. Adler, and J. Joannopoulos, *Phys. Rev. Lett.* 57, 467 (1986).

Baraff, G. A., in *Proc. 14th International Conference on Defects in Semiconductors,* ed. H. von Bardleben (Trans. Tech. Publ., Switzerland, 1986), p. 377.

Baraff, G. A., *Semicond. Sci. Technol.* 6, B9 (1991).

Baraff, G. A., in *Deep Centers in Semiconductors,* 2d Ed., ed. S. Pantelides (Gordon and Breach, Yverdon/New York, 1992), p. 547.

Baraff, G. A., E. Kane, and M. Schluter, *Phys. Rev. B* 21, 5662 (1980).

Baraff, G. A., and M. Schluter, *Phys. Rev. Lett.* 55, 1327 (1985).

Baraff, G. A., and M. Schluter, *Phys. Rev. B* 33, 7346 (1986).

Bauer, G. H., C. E. Nebel, M. B. Schubert, G. Schumm, *Mater. Res. Soc. Symp. Proc.* 149, 485 (1989).

Bauer, G. H., G. Schumm, and C.-D. Abel, in *Proc. 11th E.C. Photovoltaic Solar Energy Conference,* Montreux, Switzerland, 1992, ed. L. Guimaraes, W. Palz, C. de Reyff, H. Kiess, and P. Helm (Harwood Academic Publ., Chur, Switzerland, 1993), p. 672.

Beck, N., N. Wyrsch, E. Sauvain, and A. Shah, *Mater. Res. Soc. Symp. Proc.* 297, 479 (1993).

Benatar, L. E., M. Grimbergen, D. Redfield, and R. H. Bube, *Proc. Mat. Res. Soc.* 219, 117 (1991).

Benatar, L. E., D. Redfield, and R. Bube, *J. Appl. Phys.* 73, 8659 (1993).

201

Bennett, J. W., T. Thio, S. E. Kabakoff, D. J. Chadi, R. A. Linke, and P. Becla, *Appl. Phys. Lett.* (in press).

Bennett, M. S., A. Catalano, K. Rajan, and R. R. Arya, in *Proc. 21st IEEE Photovoltaic Specialists Conference*, Louisville, 1990 (IEEE, New York, 1990), p. 1653.

Bennett, M. S., J. L. Newton, K. Rajan, and A. Rothwarf, *J. Appl. Phys.* 62, 3698 (1987).

Beyer, W., H. Mell, and H. Overhof, in *Amorphous and Liquid Semiconductors*, ed. W. E. Spear (CICL, Edinburgh, 1977), p. 328.

Biegelsen, D. K., and R. A. Street, *Phys. Rev. Lett.* 44, 803 (1980).

Boccara, A. C., D. Fournier, and J. Badoz, *Appl. Phys. Lett.* 36, 130 (1980).

Boccara, A. C., D. Fournier, W. B. Jackson, and N. M. Amer, *Optics Lett.* 5, 377 (1980).

Boer, K. W., *Physica* 20, 1103 (1954).

Boer, K. W., *Survey of Semiconductor Physics: Electrons and Other Particles in Bulk Semiconductors* (Van Nostrand Reinhold, New York, 1990), p. 1101.

Boer, K. W., E. Borchardt, and W. Borchardt, *Z. Physik. Chem.* 203, 145 (1954).

Boer, K. W., W. Borchardt, and S. Oberlander, *Z. Physik. Chem.* 210, 218 (1959).

Branz, H., *Phys. Rev. B* 39, 5107 (1989).

Branz, H., R. Crandall, and M. Silver, *Am. Inst. Phys. Conf. Proc.* 234, 29 (1991).

Branz, H., and P. Fedders, *Mater. Res. Soc. Symp. Proc.* 336, 129 (1994).

Branz, H., and E. Schiff, *Phys. Rev. B* 48, 8667 (1993).

Branz, H., and M. Silver, *Phys. Rev. B* 42, 7420 (1990).

Brodsky, M., and R. Title, *Phys. Rev. Lett.* 23, 581 (1969).

Brunthaler, G., and K. Kohler, *Appl. Phys. Lett.* 57, 2225 (1990).

Bube, R. H., *J. Phys. Chem. Solids* 1, 234 (1957).

Bube, R. H., *J. Chem. Phys.* 30, 266 (1959).

Bube, R. H., *Photoconductivity of Solids* (Wiley and Sons, New York, 1960; reprinted by Krieger Pub. Co., Huntington, N.Y., 1978).

Bube, R. H., *Photoelectronic Properties of Semiconductors* (Cambridge Univ. Press, Cambridge, 1992).

Bube, R. H., *J. Appl. Phys.* 74, 5138 (1993).

Bube, R. H., L. E. Benatar, and K. P. Bube, *J. Appl. Phys.* (in press).

Bube, R. H., L. E. Benatar, M. N. Grimbergen, and D. Redfield, *J. Appl. Phys.* 72, 5766 (1992).

Bube, R. H., L. E. Benatar, M. N. Grimbergen, and D. Redfield, *J. Non-Cryst. Solids* 169, 47 (1994).

Bube, R. H., L. E., Benatar, and D. Redfield, *J. Appl. Phys.* 75, 1571 (1994).

Bube, R. H., G. A. Dussel, C.-T. Ho, and L. D. Miller, *J. Appl. Phys.* 37, 21 (1966).

Bube, R., L. Echeverria, and D. Redfield, *Appl. Phys. Lett.* 57, 79 (1990).

Bube, R. H., and D. Redfield, *J. Appl. Phys.* 66, 820 (1989a).

Bube, R. H., and D. Redfield, *J. Appl. Phys.* 66, 3074 (1989b).

Bube, R. H., and D. Redfield, *J. Appl. Phys.* 71, 5246 (1992).

Bullot, J., P. Cordier, M. Gauthier, and G. Mawawa, *Philos. Mag. B* 61, 413 (1990).

Calleja, E., P. Mooney, S. Wright, and M Heiblum, *Appl. Phys. Lett.* 49, 657 (1986).

Campbell, A., and B. Streetman, *Appl. Phys. Lett.* 54, 445 (1989).

Chadi, D. J., in *Proc. II–VI Conference*, Newport, Sept. 1993.

Chadi, D. J., *Annu. Rev. Mater. Sci.* 24, 45 (1994).

Chadi, D. J., and K. Chang, *Phys. Rev. Lett.* 60, 2187 (1988).
Chadi, D. J., and K. Chang, *Appl. Phys. Lett.* 55, 575 (1989a).
Chadi, D. J., and K. Chang, *Phys. Rev. B* 39, 10063 (1989b).
Chadi, D. J., and S. Zhang, *J. Electron. Mater.* 20, 55 (1991).
Chandra Sekhar, R., K. N. Raju, C. Raja Reddy, and B. K. Reddy, *J. Phys. D* 21, 1182 (1988).
Chaussat, C., E. Bustarret, J. C. Bruere, and R. Groleau, *Physica B* 129, 215 (1985).
Chen, L., and L. Yang, *J. Non-Cryst. Solids* 137/138, 1185 (1991).
Chittick, R., J. Alexander, and H. Sterling, *J. Electrochem. Soc.* 116, 77 (1969).
Cho, H., E. Kim, S. Min, K. Chang, and C Lee, *Appl. Phys. Lett.* 58, 1866 (1991).
Cody, G. D., in *Semiconductors and Semimetals,* vol. 21, ed. J. I. Pankove (Academic Press, Orlando, 1984), pt. B, p. 11.
Cohen, J. D., *J. Non-Cryst. Solids* 114, 381 (1989).
Cohen, J. D., A. V. Gelatos, K. K. Mahavadi, and K. Zellama, *Solar Cells* 24, 287 (1988).
Cohen, J. D., D. V. Lang, J. P. Harbison, and A. M. Sergent, *J. Physique (Paris)* 42, C4-371 (1981).
Cohen, J. D., T. Leen, and R. Rasmussen, *Phys. Rev. Lett.* 69, 3358 (1992).
Cohen, J., T. Leen, and F. Zhong, *J. Non-Cryst. Solids* 164/166, 327 (1993).
Craford, M., G. Stillman, J. Ross, and N. Holonyak, *Phys. Rev.* 168, 867 (1968).
Crandall, R. S., *Phys. Rev. B* 24, 7457 (1981).
Crandall, R. S., *Phys. Rev. B* 36, 2645 (1987).
Crandall, R. S., *Phys. Rev. B* 43, 4057 (1991).
Crowder, M. S., E. D. Tober, and J. Kanicki, *Appl. Phys. Lett.* 57, 1995 (1990).
Curtins, H., M. Favre, Y. Ziegler, N. Wyrsch, and A. V. Shah, *Mater. Res. Soc. Symp. Proc.* 118, 159 (1988).
Dabrowski, J., and M. Scheffler, *Phys. Rev. Lett.* 60, 2183 (1988).
Dabrowski, J., and M. Scheffler, *Phys. Rev. B* 40, 10391 (1989).
Deane, S., and M. Powell, *J. Non-Cryst. Solids* 164/166, 323 (1993a).
Deane, S., and M. Powell, *Phys. Rev. Lett.* 70, 1654 (1993b).
Debbag, F., G. Bastide, and M. Rouzeyre, *Solid State Commun.* 67, 1 (1988).
Delahoy, A., and T. Tonon, *Am. Inst. Phys. Conf. Proc.* 157, 263 (1987).
deNeufville, J. P., in *Optical Properties of Solids, New Developments,* ed. B. O. Seraphin (North-Holland, Amsterdam, 1976), p. 437.
Dersch, H., L. Schweitzer, and J. Stuke, *Phys. Rev. B* 28, 4678 (1983).
Dersch, H., J. Stuke, and J. Beichler, *Appl. Phys. Lett.* 38, 456 (1981a).
Dersch, H., J. Stuke, and J. Beichler, *Phys. Status Solidi B* 105, 265 (1981b).
Dersch, H., J. Stuke, and J. Beichler, *Phys. Status Solidi B* 107, 307 (1981c).
Dissanayake, A. S., M. Elahi, H. X. Jiang, and J. Y. Lin, *Phys. Rev. B* 45, 13996 (1992).
Dissanayake, A. S., S. X. Huang, H. X. Jiang, and J. Y. Lin, *Phys. Rev. B* 44, 13343 (1991).
Dissanayake, A. S., and H. X. Jiang, *Appl. Phys. Lett.* 61, 2048 (1993).
Dissanayake, A. S., J. Y. Lin, and H. X. Jiang, *Phys. Rev. B* 48, 8145 (1993).
Diwan, A., V. Singh, B. Arora, and P. Murawala, *J. Phys. C* 20, 3603 (1987).
Dmowski, L., S. Porowski, and M. Baj, in *High Pressure and Low Temperature Physics,* ed. C. W. Chu and J. A. Woollam (Plenum Press, New York, 1978), p. 505.
Dobaczewski, L., and P. Kaczor, *Phys. Rev. Lett.* 66, 68 (1991).
Dow, J., O. Sankey, and R. Kasoowski, *Phys. Rev. B* 43, 4396 (1991).

204   *References*

Ducharme, S., J. Hautala, and P. C. Taylor, *Phys. Rev. B* 41, 12250 (1990).
Eberhardt, K., M. Heintze, and G. H. Bauer, *J. Non-Cryst. Solids* 137/138, 187 (1991).
Ebersberger, B., W. Kruehler, and W. Fuhs, in *Proc. 11th E.C. Photovoltaic Solar Energy Conference,* Montreux, Switzerland, 1992, ed. L. Guimaraes, W. Palz, C. de Reyff, H. Kiess, and P. Helm (Harwood Academic Publ., Chur, Switzerland, 1993), p. 598.
Elliott, S. R., *Physics of Amorphous Materials* (Longman, New York, 1983).
Emin, D., in *Amorphous and Liquid Semiconductors,* ed. W. E. Spear (Univ. Edinburgh Press, Edinburgh, 1977), p. 261.
Fahrenbruch, A. L., and R. H. Bube, *J. Appl. Phys.* 45, 1264 (1974).
Fathallah, M., *Philos. Mag. B* 61, 403 (1990).
Feinleib, J., S. Groves, W. Paul, and R. Zallen, *Phys. Rev.* 131, 2070 (1963).
Fischer, D. W., *Appl. Phys. Lett.* 50, 1751 (1987).
Fischer, R., W. Rehm, J. Stuke, and U. Voget-Grote, *J. Non-Cryst. Solids* 35/36, 687 (1980).
Foyt, A. G., R. E. Halsted, and W. Paul, *Phys. Rev. Lett.* 16(2), 55 (1966).
Fritzsche, Hellmut, *Amorphous Silicon and Related Materials,* 2 vols. (World Scientific, Singapore, 1989).
Fujisawa, T., J. Yoshino, and H. Kukimoto, *Jpn. J. Appl. Phys.* 29, 388 (1990).
Gangopadhyay, S., B. Schroeder, and J. Geiger, *Philos. Mag. B* 56, 321 (1987).
Ganguly, G., and A. Matsuda, *Phys. Rev. B* 47, 3661 (1993).
Gelatos, A., and J. Kanicki, *Appl. Phys. Lett.* 57, 1197 (1990).
Gibart, P., D. L. Williamson, J. Moser, and P. Basmaji, *Phys. Rev. Lett.* 65, 1144 (1990).
Gleskova, H., P. Morin, and S. Wagner, *Appl. Phys. Lett.* 62, 2063 (1993a).
Gleskova, H., P. Morin, and S. Wagner, *Mater. Res. Soc. Symp. Proc.* 297, 589 (1993b).
Gold, R., and L. Weisberg, *Solid State Electron.* 7, 811 (1964).
Gooch, C. H., C. Hilsum, and R. B. Holeman, *J. Appl. Phys.* 32, 2069 (1961).
Goodman, N., and H. Fritzsche, *Philos. Mag. B*42, 149 (1980).
Graeff, C., R. Buhleier, and M. Stutzmann, *Appl. Phys. Lett.* 62, 3001 (1993).
Grimbergen, M., A. Lopez-Otero, A. Fahrenbruch, L. Benatar, D. Redfield, R. Bube, and R. McConville, *Mater. Res. Soc. Symp. Proc.* 258, 443 (1992).
Grimbergen, M., R. McConville, D. Redfield, and R. H. Bube, *Mater. Res. Soc. Symp. Proc.* 297, 655 (1993).
Grimmeiss, H. G., and L.-A. Ledebo, *J. Appl. Phys.* 46, 2155 (1975).
Guha, S., J. dePuydt, M. Haase, J. Qiu, and H. Cheng, *Appl. Phys. Lett.* 63, 3107 (1993).
Guha, S., and M. Hack, *J. Appl. Phys.* 58, 1683 (1985).
Gunes, M., R. M. Dawson, S. Lee, C. R. Wronski., N. Maley, and Y. M. Li, in *Proc. 22nd IEEE Photovoltaic Specialists Conference,* Las Vegas, 1991 (IEEE, New York, 1991), p. 1242.
Gunes, M., and C. R. Wronski, *Appl. Phys. Lett.* 61, 678 (1992).
Guttman, L., W. Ching, and J. Rath, *Phys. Rev. Lett.* 44, 1513 (1980).
Haase, M., P. Baude, M. Hagedorn, J. Qiu, J. dePuydt, H. Cheng, S. Guha, G. Hofler, and B. Wu, *Appl. Phys. Lett.* 63, 2315 (1993).
Hack, M., R. A. Street, and M. Shur, *J. Non-Cryst. Solids* 97/98, 803 (1987).
Haisty, R. W., E. W. Mehal, and R. Stratton, *J. Phys. Chem. Solids* 23, 829 (1962).
Hamed, A. J., *Phys. Rev. B* 44, 5585 (1991).
Hamed, A. J., and H. Fritzsche, *J. Non-Cryst. Solids* 114, 717 (1989).

Hari, P., P. Taylor, and R. Street, *J. Non-Cryst. Solids*, 164/166, 313 (1993).
Hari, P., P. Taylor, and R. Street, *Mater. Res. Soc. Symp. Proc.* 336, 329 (1994).
Hata, N., G. Ganguly, S. Wagner, and A. Matsuda, *Appl. Phys. Lett.* 61, 1817 (1992).
Hata, N., and A. Matsuda, *Appl. Phys. Lett.* 63, 1948 (1993).
Hattori, K., S. Fukuda, K. Nishimura, H. Okamoto, and Y. Hamakawa, *J. Non-Cryst. Solids* 164/166, 351 (1993).
Hauschildt, D., W. Fuhs, and H. Mell, *Phys. Status Solidi* 111, 171 (1982).
Hautala, J., W. D. Ohlsen, and P. C. Taylor, *Phys. Rev. B* 38, 11048 (1988).
Henning, J., and J. Ansems, *Semicond. Sci. Technol.* 2, 1 (1987).
Henry, C. H., and D. V. Lang, *Phys. Rev. B* 15, 989 (1977).
Hepburn, A., J. Marshall, C. Main, M. Powell, and C. van Berkel, *Phys. Rev. Lett.* 56, 2215 (1986).
Holm, B., K. Neilsen, and B. Neilsen, *Phys. Rev. Lett.* 66, 2360 (1991).
Hyun, K.-S., C. Lee, K. J. Chang, and J. Jang, *Philos. Mag. B* 64, 689 (1991).
Im, H. B., H. E. Matthews, and R. H. Bube, *J. Appl. Phys.* 41, 2581 (1970).
International Conference on Amorphous Semiconductors, J. Non-Cryst. Solids 164/166 (1993).
Iseler, G. W., J. A. Kafalas, A. J. Strauss, H. F. MacMillan, and R. H. Bube, *Solid State Commun.* 10, 619 (1972).
Isomura, M., and S. Wagner, *Mater. Res. Soc. Symp. Proc.* 258, 473 (1992).
Jackson, W. B., *Mater. Res. Soc. Symp. Proc.* 149, 571 (1989).
Jackson, W. B., and N. M. Amer, *Am. Inst. Phys. Conf. Proc.* 73, 263 (1981a).
Jackson, W. B., and N. M. Amer, *J. Phys. Colloq. Orsay Fr.* 42, C4-293 (1981b).
Jackson, W. B., and N. Amer, *Phys. Rev. B* 25, 5559 (1982).
Jackson, W. B., N. M. Amer, A. C. Boccara, and D. Fournier, *Appl. Opt.* 20, 1333 (1981).
Jackson, W. B., N. M. Johnson, and D. K. Biegelson, *Appl. Phys. Lett.* 43, 195 (1983).
Jackson, W. B., and J. Kakalios, *Phys. Rev. B* 37, 1020 (1988).
Jan, Z. S., J. C. Knights, and R. H. Bube, *J. Electron. Mater.* 8, 47 (1979).
Jan, Z. S., J. C. Knights, and R. H. Bube, *J. Appl. Phys.* 51, 3278 (1980).
Jiang, H. X., and J. Y. Lin, *Phys. Rev. Lett.* 64, 2547 (1990).
Johnson, N. M., *Appl. Phys. Lett.* 42, 981 (1983).
Johnson, N. M., D. K. Biegelsen, and M. D. Moyer, *Appl. Phys. Lett.* 40, 882 (1982).
Jones, D. P., J. Thomas, and W. A. Phillips, *Philos. Mag. B* 38, 271 (1978).
Jousse, D., and J. Kanicki, *Appl. Phys. Lett.* 55, 1112 (1989).
Kakalios, J., *J. Non-Cryst. Solids* 114, 714 (1989).
Kakalios, J., R. A. Street, and W. B. Jackson, *Phys. Rev. Lett.* 59, 1037 (1987).
Kamina, T. I., and P. J. Marcoux, *IEEE Electron. Dev. Lett.* 1, 159 (1980).
Kanev, S. K., A. L. Fahrenbruch, and R. H. Bube, *Appl. Phys. Lett.* 129, 459 (1971).
Kanev, S. K., V. Sekerdzijski, and V. Stojanov, *C. R. Acad. Bulg. Sci.* 16, 7 (1963).
Kanev, S. K., V. Stojanov, and M. Lakova, *C. R. Acad. Bulg. Sci.* 22, 863 (1969).
Kanev, S. K., V. Stojanov, and V. Sekerdzijski, *Acta Phys. Polonica* 25, 3 (1964).
Kanicki, J., ed., *Amorphous and Microcrystalline Semiconductor Devices*, 2 vols. (Artech, Boston, 1992).
Kanicki, J., and M. Sankaran, *Mater. Res. Soc. Symp. Proc.* 192, 731 (1990).
Kanicki, J., M. Sankaran, A. Gelatos, M. S. Crowder, and E. D. Tober, *Appl. Phys. Lett.* 57, 698 (1990).

Kanicki, J., W. L. Warren, C. H. Seager, M. S. Crowder, and P. M. Lenahan, *J. Non-Cryst. Solids* 137/138, 291 (1991).
Karg, F. H., H. Boehm, and K. Pierz. *J. Non-Cryst. Solids* 114, 477 (1989).
Kastner, M., D. Adler, and H. Fritzsche, *Phys. Rev. Lett.* 37, 1504 (1975).
Kastner, M. A., G. A. Thomas, and S. R. Ovshinsky, eds., *Disordered Semiconductors* (Plenum Press, New York, 1987).
Katalksy, A., and I. C. M. Hwang, *Solid State Commun.* 51, 317 (1984).
Kemp, M., and M. Silver, *Appl. Phys. Lett.* 62, 1487 (1993).
Khachaturyan, K., M. Kaminska, and E. R. Weber, *Phys. Rev. B* 40, 6304 (1989).
Kimerling, L. C., *Solid State Commun.* 16, 171 (1975).
Kimerling, L. C., *Solid State Electron.* 21, 1391 (1978).
Kimerling, L. C., in *Point and Extended Defects in Semiconductors*, ed. G. Benedek, A. Cavallani, and W. Schroter (Plenum, New York, 1988), p. 1.
Kleider, J. P., C. Longeaud, and O. Glodt, *J. Non-Cryst. Solids* 137/138, 447 (1991).
Kobayashi, K., Y. Uchida, and H. Nakashima, *Jpn. J. Appl. Phys.* 24, 928 (1985).
Kocka, J., *J. Non-Cryst. Solids* 90, 91 (1987).
Kocka, J., C. E. Nebel, and C. D. Abel, *Philos. Mag. B* 63, 221 (1991).
Kocka, J., O. Stika, and Klima, O., *Appl. Phys. Lett.* 62, 1082 (1993).
Korsunskaya, N. E., I. V. Markevich, T. V. Torchinskaya, and M. K. Sheinkman, *Phys. Status Solidi* 60, 565 (1980).
Krick, D. T., P. M. Lenahan, and J. Kanicki, *Appl. Phys. Lett.* 51, 608 (1987).
Kuech, T., E. Veuhoff, and B. Meyerson, *J. Cryst. Growth* 68, 48 (1984).
Kumeda, M., H. Yokomichi, and T. Shimizu, *Jpn. J. Appl. Phys.* 23, L502 (1984).
Landsberg, P. T., ed., *Heavy Doping and the Metal–Insulator Transition in Semiconductors, Solid State Electron.* 28(1/2) (1985).
Laks, D., C. Van de Walle, G. Neumark, and S. Pantelides, *Phys. Rev. Lett.* 66, 648 (1991).
Laks, D., C. Van de Walle, G. Neumark, and S. Pantelides, *Appl. Phys. Lett.* 63, 1375 (1993).
Lang, D. V., *Annu. Rev. Mater. Sci.* 12, 377 (1982).
Lang, D. V., in *Deep Centers in Semiconductors,* 2d Ed., ed. S. Pantelides (Gordon and Breach, Yverdon/New York, 1992), p. 591.
Lang, D. V., J. D. Cohen, and J. P. Harbison, *Phys. Rev. B* 25, 5285 (1982).
Lang, D. V., and R. Logan, *Phys. Rev. Lett.* 39, 635 (1977).
Lang, D. V., R. Logan, and M. Jaros, *Phys. Rev. B* 19, 1015 (1979).
Langer, J. M., in *New Developments in Semiconductor Physics*, ed. F. Beleznay, G. Ferenczi, and J. Giber (Springer–Verlag, Berlin 1980), p. 123.
Lannoo, M., *Semicond. Sci. Technol.* 6, B17 (1991).
Lax, M., *Phys. Rev.* 119, 1502 (1960).
LeComber, P. G., and W. E. Spear, in *Amorphous Semiconductors,* vol. 36 of *Topics in Applied Physics,* ed. M. H. Brodsky (Springer, New York, 1986), p. 251.
Lee, C., W. Ohlsen, and P. C. Taylor, *Phys. Rev. B* 31, 100 (1985).
Lee, J.-K., and E. Schiff, *Phys. Rev. Lett.* 68, 2972 (1992).
Lee, S., M. Gunes, C. R. Wronski, N. Maleyu, and M. Bennett, *Appl. Phys. Lett.* 59, 1578 (1991).
Legros, R., Y. Marfaing, and R. Triboulet, *J. Phys. Chem. Solids* 39, 79 (1978).
Lenahan, P. M., S. E. Curry, D. T. Krick, W. L. Warren, and J. Kanicki, *J. Non-Cryst. Solids* 137/138 (1991).
Levinson, M., M. Stavola, P. Besomi, and W. Bonner, *Phys. Rev. B* 30, 5817 (1984).

Ley, L., in *Hydrogenated Amorphous Silicon II,* ed. J. Joannopoulos and
G. Lucovsky (Springer–Verlag, Berlin, 1984), p. 61.

Ley, L., J. Reichardt, and R. Johnson, *Phys. Rev. Lett.* 49, 1664 (1982).

Li, Y. J., R. M. Dawson, R. W. Collins, C. R. Wronski, and S. Wiedeman, *Appl. Phys. Lett.* 59, 2459 (1991).

Lin, A. L., E. Omelianovski, and R. H. Bube, *J. Appl. Phys.* 47, 1852 (1976).

Lin, J. Y., A. S. Dissanayake, and H. X. Jiang, *Solid State Commun.* 87, 787 (1993).

Linke, R. A., T. Thio, J. D. Chadi, and G. E. Devlin, *Appl. Phys. Lett.* 65, 16 (1994).

Litton, C., and D. Reynolds, *Phys. Rev.* 125, 516 (1962).

Liu, J. Z., *Philos. Mag. Lett.* 66, 85 (1992).

Liu, J. Z., J. P. Conde, G. Lewen, and P. Roca i Cabarrocas, *Mater. Res. Soc. Symp. Proc.* 258, 795 (1992).

Liu, J. Z., G. Lewen, J. P. Conde, and P. Roca i Cabarrocas, *Mater. Res. Soc. Symp. Proc.* 297, 473 (1993).

Liu, J. Z., and P. C. Taylor, in *Amorphous and Liquid Semiconductors,* ed. F. Evangelisti and J. Stuke (North-Holland, Amsterdam, 1985), p. 1195.

Liu, J. Z., and P. C. Taylor, *Phys. Rev. B* 41, 3163 (1990).

Longeaud, C., and J. P. Kleider, *Phys. Rev. B* 45, 11672 (1992).

Lorenz, M. R., B. Segall, and H. H. Woodbury, *Phys. Rev.* 134, A751 (1964).

Losee, D. L., R. P. Khosla, D. K. Ranadive, and F. T. J. Smith, *Solid State Commun.* 13, 819 (1973).

Lowe, A. J., M. J. Powell, and S. R. Elliott, *J. Appl. Phys.* 59, 1251 (1986).

Lucovsky, G., *Solid State Commun.* 3, 299 (1965).

Luft, W., B. von Roedern, B. Stafford, D. Waddington, and L. Mrig, in *Proc. 22nd IEEE Photovoltaic Specialtists Conference,* Las Vegas, 1991 (IEEE, New York, 1991), p. 1393.

McCarthy, M., and J. Reimer, *Phys. Rev. B* 36, 4525 (1987).

MacDonald, R. L., R. A. Linke, J. D. Chadi, T. Thio, G. E. Devlin, and P. Becla, *Optics Letters* 19, 2131 (1994).

Mackenzie, K. D., P. G. LeComber, and W. E. Spear, *Philos. Mag.* 46, 377 (1982).

McMahon, T. J., *Mater. Res. Soc. Symp. Proc.* 258, 325 (1992).

McMahon, T. J., and J. P. Xi, *Phys. Rev. B* 34, 2475 (1986).

MacMillan, H. F., "Photoelectric Properties of Impurity-Associated Donors in CdTe Crystals," Ph.D. thesis, Department of Materials Science and Engineering, Stanford Univ. (1972).

Madan, A., P. G. LeComber, and W. E. Spear, *J. Non-Cryst. Solids* 20, 239 (1976).

Mandel, G., F. Morehead, and P. Wagner, *Phys. Rev.* 136, A826 (1964).

Marenko, L., V. Markwich, and L. Murin, *Sov. Phys. Semicond.* 18, 109 (1985).

Marshall, J. M., W. Pickin, A. R. Hepburn, C. Main, and R. Brueggemann, *J. Non-Cryst. Solids* 137/138, 343 (1991).

Marshall, J. M., R. A. Street, and M. J. Thomson, *Phys. Rev. B* 29, 2331 (1984).

Martin, D., B. Schroeder, M. Leidner, and H. Oechsner, *J. Non-Cryst. Solids* 114, 537 (1989).

Martin, G. M., *Appl. Phys. Lett.* 39, 747 (1981).

Materials Research Society, *Amorphous Silicon Technology 1994, Mater. Res. Soc. Symp. Proc.* 336 (1994).

Meaudre, M., and R. Meaudre, *Phys. Rev.* 45, 4524 (1992a).

Meaudre, R., and M. Meaudre, *Phys. Rev.* 45, 12134 (1992b).

Meaudre, R., S. Vignoli, and M. Meaudre, *Philos. Mag. Lett.* 69, 327 (1994).

Mell, H., and W. Beyer, *J. Non-Cryst. Solids* 59/60, 405 (1983).
Mitonneau, A., and A. Mircea, *Solid State Commun.* 30, 157 (1979).
Mizuta, M., M. Tachikawa, H. Kukimoto, and S. Minomura, *Jpn. J. Appl. Phys.* 24, L143 (1985).
Moddel, G., D. A. Anderson, and W. Paul, *Phys. Rev. B* 22, 1918 (1980).
Mooney, P. M., *J. Appl. Phys.* 67, R1 (1990).
Mooney, P. M., *Semicond. Sci. Technol.* 6, B1 (1991).
Mooney, P. M., T. Theis, and E. Calleja, *J. Electron. Mater.*, 20, 23 (1991).
Mooney, P. M., T. Theis, and S. Wright, *Appl. Phys. Lett.* 53, 2546 (1988).
Morgan, T. N., in *Defects in Semiconductors 15*, ed. G. Ferenczi, *Mater. Sci. Forum*, vols. 38–41 (Trans. Tech. Publ., Switzerland, 1989), p. 1079.
Morgan, T. N., *Semicond. Sci. Technol.* 6, B23 (1991).
Morigaki, K., in *Semiconductors and Semimetals*, vol. 21, ed. J. I. Pankove (Academic Press, Orlando, 1984), pt. C, p. 155.
Morigaki, K., I. Hirabayashi, M. Nakayama, S. Nitta, and K. Shimikawa, *Solid State Commun.* 33, 851 (1980).
Morimoto, A., M. Matsumoto, M. Yoshita, M. Kumeda, and T. Shimizu, *Appl. Phys. Lett.* 59, 2130 (1991).
Morimoto, A., Y. Tsujimura, M. Kumeda, and T. Shimizu, *Jpn. J. Appl. Phys.* 24, 1394 (1985).
Mott, N. F., and E. A. Davis, *Electronic Processes in Non-Crystalline Materials*, 2d Ed. (Oxford Univ. Press, Oxford, 1979).
Mott, N. F., E. A. Davis, and R. A. Street, *Philos. Mag.* 32, 961 (1975).
Nakamara, G., K. Sato, and Y. Yukimoto, *Solar Cells* 9, 75 (1983).
Nakata, M., S. Wagner, and T. Peterson, *J. Non-Cryst. Solids* 164/166, 179 (1993).
Nebel, C. E., G. H. Bauer, M. Gorn, and P. Lechner, in *Proc. 9th European Photovoltaic Conference* (Kluwer Academic Publ., Dordrecht, 1989), p. 267.
Nelson, R. J., *Appl. Phys. Lett.* 31, 351 (1977).
Neumark, G. F., in *Wide-Gap II–VI Compounds for Opto-electronic Applications*, ed. H. Ruda (Chapman & Hall, London, 1992).
Nevin, W. A., H. Yamagishi, and Y. Tawada, *Appl. Phys. Lett.* 54, 1226 (1989).
Ngai, K. L., T. L. Reinecke, and E. N. Enconomou, *Phys. Rev. B* 17, 790 (1978).
Ngai, K. L., and P. C. Taylor, *Philos. Mag. B* 37, 175 (1978).
Nicholas, K. H., and J. Woods, *Br. J. Appl. Phys.* 16, 783 (1964).
Nickel, N. H., W. B. Jackson, and N. M. Johnson, *Phys. Rev. Lett.* 71, 2733 (1993).
Nishizawa, J., Y. Oyama, and K. Dezaki, *J. Appl. Phys.* 75, 4482 (1994).
Norberg, R., J. Bodart, R. Corey, P. Fedders, W. Paul, W. Turner, D. Pang, and A. Wetsel, *Mater. Res. Soc. Symp. Proc.* 258, 377 (1992).
Northrop, G., and P. Mooney, *J. Electron. Mater.* 20, 13 (1991).
Oheda, H., *J. Appl. Phys.* 52, 6693 (1981).
Overhof, H., *Mater. Res. Soc. Symp. Proc.* 258, 681 (1992).
Overhof, H., and P. Thomas, *Hydrogenated Amorphous Semiconductors* (Springer–Verlag, Berlin, 1989).
Ovshinsky, S. R., *Phys. Rev. Lett.* 36, 1469 (1976).
Pandya, R., and E. A. Schiff, *J. Non-Cryst. Solids* 77/78, 623 (1985).
Pankove, J. I., *Semiconductors and Semimetals*, vol. 21 (pts. A–D) (Academic Press, Orlando, 1984).
Pankove, J. I., and J. Berkeyheiser, *Appl. Phys. Lett.* 37, 705 (1980).
Park, H., J. Liu, and S. Wagner, *Appl. Phys. Lett.* 55, 2658 (1989).

Park, H., J. Liu, P. Roca i Cabaroccas, A. Maruyama, M. Isomura, S. Wagner, J. Abelson, and F. Finger, *Appl. Phys. Lett.* 57, 1440 (1990a).

Park, R., M. Troffer, C. Rouleau, J. DePuydt, and M. Haase, *Appl. Phys. Lett.* 57, 2127 (1990b).

Parker, M. A., K. A. Conrad, and E. A. Schiff, *Mater. Res. Soc. Symp. Proc.* 70, 125 (1986).

Parker, M. A., and E. A. Schiff, *Appl. Phys. Lett.* 48, 1087 (1986).

Pastor, K., and R. Triboulet, in *Proc. 18th International Conference on the Physics of Semiconductors,* Stockholm, Sweden, 1986, ed. O. Engstrom (World Scientific, Singapore, 1987), vol. 2, p. 859.

Paul, W., in *Proc. Ninth International Conference on the Physics of Semiconductors* (Nauka, Leningrad, 1968), vol. 2, p. 16.

Paul, W., *J. Non-Cryst. Solids* 137/138, 803 (1991).

Petroff, P., and R. Hartman, *Appl. Phys. Lett.* 23, 469 (1973).

Platz, R., R. Brueggemann, and G. H. Bauer, *J. Non-Cryst. Solids* 164/166, 355 (1993).

Plonka, A., *Time-Dependent Reactivity of Species in Condensed Media* (Springer–Verlag, Berlin, 1986).

Powell, M. J., *Appl. Phys. Lett.* 43, 597 (1983).

Powell, M. J., and D. Nicholls, *Proc. IEEE* 130, 2 (1983).

Powell, M. J., C. van Berkel, and S. Deane, *J. Non-Cryst. Solids* 137/138, 1215 (1991).

Powell, M. J., C. van Berkel, I. French, and D. Nicholls, *Appl. Phys. Lett.* 51, 1242 (1987).

Press, W. H., B. P. Flannery, S. A. Teukolsky, and W. T. Vetterling, *Numerical Recipes in PASCAL* (Cambridge Univ. Press, Cambridge, 1989).

Queisser, H. J., *Phys. Rev. Lett.* 54, 234 (1985).

Queisser, H. J., and D. E. Theodorou, *Phys. Rev. Lett.* 43, 401 (1979).

Queisser, H. J., and D. E. Theodorou, *Phys. Rev. B* 33, 4027 (1986).

Rath, J. K., B. Hackenbuchner, W. Fuhs, and H. Mell, in *Amorphous Silicon Materials and Solar Cells*, ed. B. Stafford, *Am. Inst. Phys. Conf. Proc.* 234 (AIP, New York, 1991), p. 98.

Redfield, D., *Appl. Phys. Lett.* 48, 846 (1986a).

Redfield, D., *Appl. Phys. Lett.* 49, 1517 (1986b).

Redfield, D., *Appl. Phys. Lett.* 52, 492 (1988).

Redfield, D., *Appl. Phys. Lett.* 54, 398 (1989).

Redfield, D., *Mod. Phys. Lett. B* 5, 933 (1991).

Redfield, D., *Mater. Res. Soc. Symp. Proc.* 258, 341 (1992).

Redfield, D., *Bull. Amer. Phys. Soc.* 40, 255 (1995).

Redfield, D., and R. H. Bube, *Appl. Phys. Lett.* 54, 1037 (1989).

Redfield, D., and R. H. Bube, *Phys. Rev. Lett.* 65, 464 (1990).

Redfield, D., and R. H. Bube, *J. Non-Cryst. Solids* 137/138, 215 (1991a).

Redfield, D., and R. H. Bube, *Mater. Res. Soc. Symp. Proc.* 219, 21 (1991b).

Redfield, D., and R. H. Bube, in *Proc. 22nd IEEE Photovoltaic Specialists Conference,* Las Vegas, 1991 (IEEE, New York, 1991c), p. 1319.

Redfield, D., and R. H. Bube, *Proc. Mater. Res. Soc. Symp. Proc.* 297, 607 (1993).

Redfield, D., and R. H. Bube, *Mater. Res. Soc. Symp. Proc.* 336, 263 (1994).

Rekhson, S., in *Glass: Science and Technology,* vol. 3, ed. D. Uhlmann and N. Kreidl (Academic Press, New York, 1986), p. 1.

Rossi, M. C., M. S. Brandt, and M. Stutzmann, *Appl. Phys. Lett.* 60, 1709 (1992).

Sakata, I., M. Yamanaka, S. Numase, and Y. Hayashi, *J. Appl. Phys.* 71, 4344 (1992).

Saleh, Z. M., H. Tarui, N. Nakamura, M. Nishikuni, S. Tsuda, S. Nakano, and Y. Kuwano, *Jpn. J. Appl. Phys.* 12, 3801 (1992).
Saleh, Z. M., H. Tarui, S. Tsuda, S. Nakano, and Y. Kuwano, *Mater. Res. Soc. Symp. Proc.* 297, 501 (1993).
Santos, P. V., C. F. deO. Graeff, I. Chambouleyron, *J. Non-Cryst. Solids* 128, 243 (1991).
Schade, H., in *Semiconductors and Semimetals,* vol. 21, ed. J. I. Pankove (Academic Press, Orlando, 1984), pt. B, p. 359.
Scheffler, M., in *Festkorperprobleme-29,* ed. U. Roessler (Vieweg, Braunschweig, 1989), p. 231.
Scherer, G. W., *Relaxation in Glasses and Composites* (John Wiley, New York, 1986).
Schiff, E. A., *Philos. Mag. Lett.* 55, 87 (1987a).
Schiff, E. A., in *Disordered Semiconductors,* ed. M. A. Kastner, G. A. Thomas, S. R. Ovshinsky (Plenum Press, New York, 1987b), p. 379.
Schneider, U., and B. Schroeder, in *Amorphous Silicon and Related Materials,* ed. H. Fritzsche (World Scientific, Singapore, 1989a), vol. A, p. 687.
Schneider, U., and B. Schroeder, *Solid State Commun.* 69, 895 (1989b).
Schneider, U., B. Schroeder, and F. Finger, *J. Non-Cryst. Solids* 97/98, 795 (1987).
Scholz, A., W. Herbst, B. Schroeder, and P. Lechner, in *Proc. 11th E.C. Photovoltaic Solar Energy Conference,* Montreux, Switzerland, 1992, ed. L. Guimaraes, W. Palz, C. de Reyff, H. Kiess, and P. Helm (Harwood Academic Publ., Chur, Switzerland, 1993), p. 754.
Scholz, A., B. Schehr, and B. Schroeder, *Solid State Commun.* 85, 753 (1993).
Scholz, A., and B. Schroeder, *Am. Inst. Phys. Conf. Proc.* 234, 178 (1990).
Scholz, A., and B. Schroeder, *J. Non-Cryst. Solids* 137/138, 259 (1991).
Scholz, A., B. Schroeder, and H. Oechsner, in *Proc. 23rd IEEE Photovoltaic Specialists Conference,* Louisville, 1993 (IEEE, New York, 1993), p. 1653.
Schropp, R., and J. Verwey, *Appl. Phys. Lett.* 50, 185 (1987).
Schumm, G., *Phys. Rev. B* 49, 2427 (1994).
Schumm, G., and G. H. Bauer, *Phys. Rev. B* 39, 5311 (1989).
Schumm, G., and G. H. Bauer, *J. Non-Cryst. Solids* 137/138, 315 (1991a).
Schumm, G., and G. H. Bauer, in *Proc. 22nd IEEE Photovoltaic Specialists Conference,* Las Vegas, 1991 (IEEE, New York, 1991b), p. 1225.
Schumm, G., K. Nitsch, and G. H. Bauer, *Philos. Mag. B* 58, 411 (1988).
Seager, C. H., and J. Kanicki, *Appl. Phys. Lett.* 57, 1378 (1990).
Semaltianos, N. G., G. Karczewski, T. Wojtowicz, and J. K. Furdyna, *Phys. Rev. B* 47, 12540 (1993).
Sheinkman, M. K., in *Proc. 18th International Conference on the Physics of Semiconductors,* Stockholm, Sweden, 1986, ed. O. Engstrom (World Scientific, Singapore, 1987), vol. 2, p. 785.
Sheinkman, M. K., et al., *Fiz. Tekh. Poluprov.* 5, 1904 (1971).
Sheinkman, M. K., N. E. Korsunskaya, I. V. Markevich, and T. V. Torchinskaya, *J. Phys. Chem. Solids* 43, 475 (1982).
Sheinkman, M. K., and A. Ya. Shik, *Fiz. Tekh. Poluprov.* 10, 209 (1976).
Shepard, K., Z. E. Smith, S. Aljishi, and S. Wagner, *Appl. Phys. Lett.* 53, 1644 (1988).
Shimakawa, K., S. Inami, and S. R. Elliott, *Phys. Rev. B* 42, 11857 (1990).
Shimakawa, K., S. Inami, and S. R. Elliott, *J. Non-Cryst. Solids* 137/138, 1017 (1991).
Shimakawa, K., A. Kolobov, and S. R. Elliott, *Advances in Physics* (in press).

Shimizu, T., H. Kidoh, M. Matsumoto, A. Morimoto, and M. Kumeda, *J. Non-Cryst. Solids* 114, 630 (1989).

Siebke, F., and H. Stiebig, *Mater. Res. Soc. Symp. Proc.* 336, 371 (1994).

Skierbiszewski, C., T. Suski, P. Wisniewski, W. Jandtsch, G. Ostermayer, Z. Wilamowski, P. Walker, N. Mason, and J. Singleton, *Appl. Phys. Lett.* 63, 3209 (1993).

Skumanich, A., and N. M. Amer, *Appl. Phys. Lett.* 52, 643 (1988).

Skumanich, A., N. Amer, and W. Jackson, *Phys. Rev. B* 31, 2263 (1985).

Smail, T., and T. Mohammed-Brahim, *Philos. Mag. B* 64, 675 (1991).

Smail, T., and T. Mohammed-Brahim, in *Proc. 11th E.C. Photovoltaic Solar Energy Conference,* Montreux, Switzerland, 1992, ed. L. Guimaraes, W. Palz, C. de Reyff, H. Kiess, and P. Helm (Harwood Academic Publ., Chur, Switzerland, 1993), p. 699.

Smith, G. B., *J. Appl. Phys.* 62, 3380 (1987).

Smith, Z, and S. Wagner, *Phys. Rev. B* 32, 5510 (1985).

Smith, Z, and S. Wagner, *Phys. Rev. Lett.* 59, 688 (1987).

Smith, Z, and S. Wagner, in *Amorphous Silicon and Related Materials,* ed. H. Fritzsche (World Scientific, Singapore, 1989), p. 409.

Snell, A., K. Mackenzie, W. Spear, P. LeComber, and A. Hughes, *Appl. Phys.* 24, 357 (1981).

Spear, W., and P. LeComber, *Philos. Mag.* 33, 935 (1976).

Staebler, D. L., R. Crandall, and R. Williams, *Appl. Phys. Lett.* 39, 733 (1981).

Staebler, D. L., and C. R Wronski, *Appl. Phys. Lett.* 31, 292 (1977).

Staebler, D. L., and C. R. Wronski, *J. Appl. Phys.* 51, 3262 (1980).

Stafford, B., ed., *Amorphous Silicon Materials and Solar Cells, Am. Inst. Phys. Conf. Proc.* 234 (AIP, New York, 1991).

Stafford, B., and E. Sabisky, eds., *Stability of Amorphous Silicon Alloy Materials and Devices, Am. Inst. Phys. Conf. Proc.* 157 (AIP, New York, 1987).

Stavola, M., M. Levinson, J. Benton, and L. Kimerling, *Phys. Rev. B* 30, 832 (1984).

Stoneham, A. M., *Philos. Mag. B* 51, 161 (1985).

Stradins, P., and H. Fritzsche, *Mater. Res. Soc. Symp. Proc.* 297, 571 (1993).

Stradins, P., and H. Fritzsche, *Philos. Mag.* 69, 121 (1994).

Street, R. A., *Appl. Phys. Lett.* 41, 1060 (1982a).

Street, R. A., *Phys. Rev. Lett.* 49, 1187 (1982b).

Street, R. A., *Appl. Phys. Lett.* 42, 507 (1983).

Street, R. A., in *Semiconductors and Semimetals,* vol. 21, ed. J. I. Pankove (Academic Press, New York, 1984), pt. B, p. 197.

Street, R. A., *J. Non-Cryst. Solids* 77/78, 1 (1985).

Street, R. A., *Mater. Res. Soc. Symp. Proc.* 95, 13 (1987).

Street, R. A., *Appl. Phys. Lett.* 59, 1084 (1991a).

Street, R. A., *Hydrogenated Amorphous Silicon* (Cambridge Univ. Press, Cambridge, 1991b).

Street, R. A., D. K. Biegelsen, and J. C. Knights, *Phys. Rev. B* 24, 969 (1981).

Street, R. A., D. K. Biegelsen, and J. Stuke, *Philos. Mag. B* 40, 451 (1979).

Street, R. A., and M. Hack, *J. Non-Cryst. Solids* 137/138, 263 (1991).

Street, R. A., M. Hack, and W. Jackson, *Phys. Rev. B* 37, 4209 (1988).

Street, R. A., J. Kakalios, and T. Hayes, *Phys. Rev. B* 34, 3030 (1986).

Street, R. A., and K. Winer, *Mater. Res. Soc. Symp. Proc.* 149, 131 (1989).

Stuke, J., *J. Non-Cryst. Solids* 97/98, 1 (1987).

Stutzmann, M., *Philos. Mag. B* 56, 63 (1987).

Stutzmann, M., and W. Jackson, *Solid State Commun.* 62, 153 (1987).

Stutzmann, M., W. Jackson, and C. Tsai, *Phys. Rev. B* 32, 23 (1985).
Stutzmann, M., J. Nunnenkamp. M. S. Brandt, and A. Asano, *Phys. Rev. Lett.* 67, 2347 (1991a).
Stutzmann, M., J. Nunnenkamp, M. S. Brandt, A. Asano, and M. C. Rossi, *J. Non-Cryst. Solids* 137/138, 231 (1991b).
Szafranek, I., S. Bose, and G. Stillman, *Appl. Phys. Lett.* 55, 1205 (1989).
Sze, S. M., *Physics of Semiconductor Devices* (Wiley, New York, 1969).
Takebe, T., J. Saraie, and H. Matsunami, *J. Appl. Phys.* 53, 457 (1982).
Tanaka, K., *Am. Inst. Phys. Conf. Proc.* 31, 148 (1976).
Tanaka, K., *J. Non-Cryst. Solids* 35/36, 1023 (1980a).
Tanaka, K., *Solid State Commun.* 34, 201 (1980b).
Tanaka, K., *Phys. Rev. B* 30, 4549 (1984).
Tanaka, K., *J. Non-Cryst. Solids* 137/138, 1 (1991).
Tanaka, K., and M. Kikuchi, *Solid State Commun.* 11, 1311 (1972).
Tanaka, K., and R. Tsu, *Phys. Rev. B.* 24, 2038 (1981).
Tang, Y., R. Braunstein, B. von Roedern, and F. R. Shapiro, *Mater. Res. Soc. Symp. Proc.* 297, 407 (1993).
Taylor, P. C., in *Semiconductors and Semimetals,* vol. 21, ed. J. I. Pankove (Academic Press, Orlando, 1984), pt. C, p. 99.
Taylor, P. C., in *Amorphous Semiconductors*, ed. M. Pollak (CRC Press, New York, 1987a), p. 19.
Taylor, P. C., in *Amorphous Semiconductors*, ed. M. Pollak (CRC Press, New York, 1987b), p. 69.
Taylor, P. C., in *Laser Spectroscopy of Solids II,* vol. 65 of *Topics in Applied Physics,* ed. W. M. Yen (Springer–Verlag, Berlin/Heidelberg, 1989), p. 257.
Taylor, P. C., and J. Z. Liu, in *Defects in Glasses, Mater. Res. Soc. Symp. Proc.* 61, 223 (1985).
Terakawa, A., M. Shima, K. Sayama, H. Tarui, H. Nishiwaki, and S. Tsuda, *Mater. Res. Soc. Symp. Proc.* 336, 487 (1994).
Thio, T., R. A. Linke, G. E. Devlin, J. W. Bennett, J. D. Chadi, and M. Mizuta, *Appl. Phys. Lett.* 65, 1802 (1994).
Tober, E. D., J. Kanicki, and M. S. Crowder, *Appl. Phys. Lett.* 59, 1723 (1991).
Tran, M., H. Fritzsche, and P. Stradins, *Mater. Res. Soc. Symp. Proc.* 297, 195 (1993).
Treacy, D. J., U. Strom, P. B. Klein, P. C. Taylor, and T. P. Martin, *J. Non-Cryst. Solids* 35/36, 1035 (1980).
Treacy, D. J., P. C. Taylor, and P. B. Klein, *Solid State Commun.* 32, 423 (1979).
Tsang, C., and R. Street, *Philos. Mag. B* 37, 601 (1978).
Tscholl, E., "The Photochemical Interpretation of Slow Phenomena, in Cadmium Sulphide," *Philips Res. Repts. Suppl.* No. 6 (1968).
Tsukada, N., T. Kikuta, and K. Ishida, in *Proc. International Symposium on GaAs and Related Compounds,* Kruizawa, Japan, 1985, ed. M. Fujimoto (Hilger, London, 1985).
Turner, W. A., S. J. Jones, D. Pang, B. F. Bateman, J. H. Chen, Y.-M. Li, F. C. Marques, A. E. Wetsel, P. Wickboldt, W. Paul, J. Bodat, R. E. Norbert, I. El Zawawi, and M. L. Theye, *J. Appl. Phys.* 67, 7430 (1990).
Unold, T., and J. D. Cohen, *Appl. Phys. Lett.* 58, 723 (1991).
Unold, T., J. D. Cohen, and C. M. Fortmann, *Mater. Res. Soc. Symp. Proc.* 258, 499 (1992).
Unold, T., J. Hautala, and J. D. Cohen, *Phys. Rev. B* 50, 16985 (1994).
Vaillant, F., D. Jousse, and J.-C. Bruyere, *Philos. Mag. B* 57 649 (1988).
van Berkel, C., and M. Powell, *Appl. Phys. Lett.* 51, 1094 (1987).

Van de Walle, C., D. Laks, G. Neumark, and S. Pantelides, *Phys. Rev. B* 47, 9425 (1993).

Vanecek, M., J. Fric, R. Crandall, and A. Mahan, *J. Non-Cryst. Solids* 164/166, 335 (1993).

Vanecek, M., J. Kocka, J. Stuchlik, and A. Triska, *Solid State Commun.* 39, 1199 (1981).

Vanier, P. E., *Appl. Phys. Lett.* 41, 986 (1982).

Vanier, P. E., in *Semiconductors and Semimetals*, vol. 21, ed. J. I. Pankove (Academic Press, Orlando, 1984), p. 329.

Vanier, P. E., and R. W. Griffith, *J. Appl. Phys.* 53, 3098 (1982).

Voget-Grote, U., W. Kuemmerle, R. Fischer, and J. Stuke, *Philos. Mag. B* 41, 127 (1980).

Vul, A. Ya., Sh. I. Nabiev, S. G. Petrosyan, and A. Ya. Shik, *Phys. Status Solidi A* 36, 53 (1976).

Vul, A. Ya., and A. Ya. Shik, *Solid State Commun.* 13, 1049 (1973).

Waag, A., F. Fischer, J. Gerschuetz, S. Scholl, and G. Landwehr, *J. Appl. Phys.* 75, 1368 (1994).

Wagner, P., and J. Hage, *Appl. Phys. A* 49, 123 (1989).

Walukiewicz, W., in *Proc. 17th International Conference on Defects in Semiconductors*, 1993 (Materials Science Forum, Switzerland, 1994).

Wang, S. L., Z. H. Lin, J. M. Viner, and P. C. Taylor, *Mater. Res. Soc. Symp. Proc.* 336, 559 (1994).

Wang, S. L., J. M. Viner, M. Anani, and P. C. Taylor, *J. Non-Cryst. Solids* 164/166, 251 (1993).

Warren, W. L., J. Kanicki, J. Robertson, and P. M. Lenahan, *Appl. Phys. Lett.* 59, 1699 (1991).

Warren, W. L., P. M. Lenahan, and S. E. Curry, *Phys. Rev. Lett.* 65, 207 (1990).

Warren, W. L., P. M. Lenahan, and J. Kanicki, *J. Appl. Phys.* 70, 2220 (1991).

Warren, W. L., J. Robertson, and J. Kanicki, *Appl. Phys. Lett.* 63, 2685 (1993).

Watkins, G. D., in *Proc. International Symposium on DX-Centers and Other Metastable Defects in Semiconductors, Semicond. Sci. Technol.* 6, B111 (1991).

Watkins, G. D., in *Deep Centers in Semiconductors,* 2d Ed., ed by S. Pantelides (Gordon and Breach, Yverdon/New York, 1992), p. 177.

Wiedeman, S., M. S. Bennett, and J. L. Newton, *Mater. Res. Soc. Symp. Proc.* 95, 145 (1987).

Winer, K., *Phys. Rev. B* 41, 12150 (1990).

Winer, K., *J. Non-Cryst. Solids* 137/138, 157 (1991).

Winer, K., I. Hirabayashi, and L. Ley, *Phys. Rev. B* 38, 7680 (1988).

Wolk, J. A., M. B. Kruger, J. N. Heyman, W. Walukiewicz, R. Jeanloz, and E. E. Haller, *Phys. Rev. Lett.* 66, 774 (1991).

Wolk, J. A., W. Walukiewicz, M. Thewalt, and E. Haller, *Phys. Rev. Lett.* 68, 3619 (1992).

Woods, J. F., *J. Electron. Control* 5, 417 (1958).

Woods, J. F., and N. G. Ainslie, *J. Appl. Phys.* 34, 1469 (1963).

Woods, J. F., and K. H. Nicholas, *Br. J. Appl. Phys.* 15, 1361 (1964).

Woods, J. F., and D. A. Wright, *International Conference on Solid-State Physics,* Brussels, 1958 (Academic Press, New York, 1960), p. 880.

Wright, H. C., R. J. Downey, and J. R. Canning, *Br. J. Appl. Phys. (J. Phys. D) Ser. 2,* 1, 1593 (1968).

Wu, Z., J. Siefert, and B. Equer, *J. Non-Cryst. Solids,* 137/138, 227 (1991).

Wyrsch, N., F. Finger, T. J. McMahon, and M. Vanecek, *J. Non-Cryst. Solids* 137/138, 347 (1991).

Xu, X., J. Yang, and S. Guha, *Appl. Phys. Lett.* 62, 1399 (1993).
Yamaguchi, M., K. Ando, A. Yamamoto, and C. Uemura, *Appl. Phys. Lett.* 44, 432 (1984).
Yang, L., and L Chen, *Appl. Phys. Lett.* 63, 400 (1993a).
Yang, L., and L Chen, *Mater. Res. Soc. Symp. Proc.* 297, 619 (1993b).
Yoon, J.-K., J. Jang, and C. Lee, *J. Appl. Phys.* 64, 6591 (1988).
Yoshino, J., M. Tachikawa, N. Matsuda, M. Mizuta, and H. Kukimoto, *Jpn. J. Appl. Phys.* 23, L29 (1984).
Yu, P. W., *Solid State Commun.* 43, 953 (1982).
Zallen, R., *The Physics of Amorphous Solids* (Wiley, New York, 1983).
Zallen, R., and W. Paul, *Phys. Rev.* 134 (6A), 1628 (1964).
Zhang, Q., H. Takashima, J-H. Zhou, M. Kumeda, and T. Shimizu, *Phys. Rev. B.* 50, 1551 (1994).
Zhong, F., and J. D. Cohen, *Mater. Res. Soc. Symp. Proc.* 258, 813 (1992).

Index